The Flagellar World

The Flagellar World

Electron microscopic images of bacterial flagella and related surface structures from more than 30 species

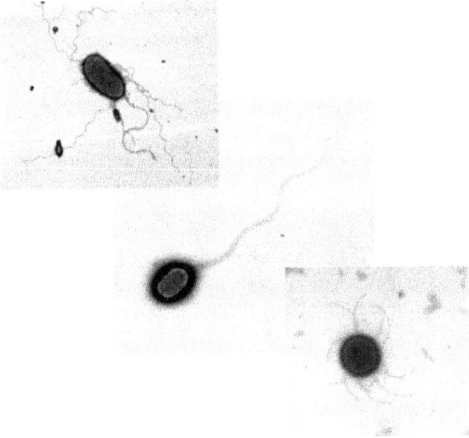

Shin-Ichi Aizawa
Prefectural University of Hiroshima

ELSEVIER

AMSTERDAM • BOSTON • HEIDELBERG • LONDON
NEW YORK • OXFORD • PARIS • SAN DIEGO
SAN FRANCISCO • SINGAPORE • SYDNEY • TOKYO
Academic Press is an imprint of Elsevier

Academic Press is an imprint of Elsevier
The Boulevard, Langford Lane, Kidlington, Oxford, OX5 1GB, UK
225 Wyman Street, Waltham, MA 02451, USA

First published 2014

British Library Cataloguing in Publication Data
A catalogue record for this book is available from the British Library

Library of Congress Cataloging-in-Publication Data
A catalog record for this book is available from the Library of Congress

ISBN: 978-0-12-417234-0

For information on all Academic Press publications
visit our website at **store.elsevier.com**

This book has been manufactured using Print On Demand technology. Each copy is produced to
order and is limited to black ink. The online version of this book will show color figures where
appropriate.

ELSEVIER | Book Aid International

Working together
to grow libraries in
developing countries

www.elsevier.com • www.bookaid.org

Transferred to Digital Printing in 2013

CONTENTS

Preface...vii
Introduction...1

Chapter examples...12
1   *Actinoplanes missouriensis* — Swimming Spores with Flagella 14
2   *Aliivibrio fischeri* — Light-Organ Symbiont in the Bobtail
    Squid 16
3   *Azospirillum brasilense* — A Bushy Hook of the Polar
    Flagellum 18
4   *Bacillus subtilis* — The Representative of Gram-Positive
    Bacteria 22
Topic 1: Gene Regulation 24
5   *Bdellovibrio bacteriovorus* — A Small but Fierce Predator.......... 26
6   *Borrelia burgdorferi* — Periplasmic Flagella in a Flat
    Wave Body 28
7   *Bradyrhizobium japonicum* — The Nitrogen–Fixer in the
    Peanut Farm 30
8   *Caulobacter crescentus* — Alteration between Flagellum and
    Stalk.. 32
9   *Enterococcus casseliflavus* — Edible Flagella............................. 34
10 *Escherichia coli* — The Representative of the Gram-Negative
    Bacteria 36
11 *Geobacillus kaustophilus* — The Heat- and Acid-Stable
    Flagella.. 40
12 *Gluconobacter oxydans* — The Vinegar Producing Bacteria......... 42
13 *Helicobacter pylori* — Randomly Arranged Flagellar Genes 44
Topic 2: Gene Arrangement 46
Topic 3: Mot Proteins... 47
14 *Idiomarina loihiensis* — A Habitat of Deep-Sea Volcano 48
15 *Legionella pneumophila* — Opportunistic Pathogen in
    Public Bath 50
16 *Magnetospirillum magnetotacticum* — High-Quality
    Magnet in the Pond 52

Topic 4: Flagellin size 54

17 *Paenibacillus alvei* — Flagella-Dependent Social Motility 56

18 *Pectobacterium carotovorum* — Subpolar Hyper-Flagellation...... 58

19 *Pseudomonas aeruginosa* — Opportunistic Pathogen
    in the Hospital.. 60

Topic 5: Flagella and Pathogenicity... 63

20 *Ralstonia solanacearum* — Ubiquitous Plant Pathogen.............. 64

21 *Rhodobacter sphaeroides* — A Resourceful Little Bug 66

Topic 6: Flagellar Position and Shape 69

22 *Ruegeria* sp. TM1040 — A Fast Swimmer in the Sea 70

23 *Saccharophagus degradans* — The Seaweed Eater 72

24 *Salmonella enterica* Serovar Typhimurium — The Best-Studied
    Flagella... 74

Topic 7: History of *Salmonella* SJW Strains..................................... 77

25 *Selenomonas ruminantium* — The Authentic Lateral Flagella 78

Topic 8: Hook length 79

Topic 9: Multiple Flagellins 80

26 *Sinorhizobium meliloti* — Nitrogen-Fixer in the Grassland 82

27 *Symbiobacterium thermophilum* — A Gram-Negative, High
    (G + C) Firmicutes.. 84

28 *Vibrio parahaemolyticus* — Polar/Lateral Flagella with
    H+/Na+ Motor 86

29 *Xanthomonas oryzae* pv. *Oryzae*— Pathogen in the
    Rice Country... 88

30 *Uncharacterized Species* — Slowly-Growing Bacteria 90

31 *Buchnera aphidicola* — Flagella Not for Motility 92

32 *Methanococcus voltae* — Archaeal Flagella or Archaellum........ 94

33 *Myxococcus xanthus* — To be Social or to be Adventurous 96

34 *Saprospira grandis* — A Grand Predator on the Seashore......... 98

35 *Shigella flexneri* — Flagellaless *E. coli* 100

Appendix 102

References 115

Index   131

I graduated from the physics department of Tohoku University in 1974 and moved into a brand-new (at the time) science, Biophysics, at Nagoya University, where I met bacterial flagella as one of the rotation programs for graduates (under Prof. Sho Asakura). As a physics student, I was very much interested in the flagellar motor. I attempted to purify the motor and failed, which gave me an opportunity to quickly learn biology, biochemistry, genetics, and molecular biology. As a postdoc (1980–1984, under the late Prof. Robert Macnab) at Yale University, I successfully developed a new method for purification of flagella with the motor still attached. Since then, flagella has been the main theme of my research.

In the early years of research, I worked with *Salmonella typhimurium* only, using abundant mutants of SJW strains (see Topic: History of *Salmonella* SJW (Salmonella Japan Waseda) strains). After 10 years of constant publications on *Salmonella* flagella, people from around the world started sending their "bugs" for me to take a look at with the electron microscope (EM). I accepted almost all bugs. I was very happy to see their flagella, even though some of them were human pathogens.

Now I am bewildered by the piles of electron microscope negatives, as my retirement is getting close. I took thousands of EM pictures of flagella from various species. Some of them were published, but many were left unused. What should I do with them: burn them or shred them? Rather than terminating them forever, I chose a way to leave them in a book. This book, then, is naturally not a review of flagella research of the world, but a collection of my work on flagella with my view of this small world.

This book contains 35 strains of eubacteria and archaeal species, aligned one by one in alphabetical order. Each chapter contains one species on two facing pages. On the first page, I show pictures of a whole cell, then isolated flagella, and then other cell surface appendages, if any, and on the second page is the genetic map regarding flagellar genes and the analysis of genes. Five species were placed into an exceptions section, because they are not really flagella. In the introduction, I summarize the flagellar structures and the flagellar genetics for those who are not familiar with this world. In the appendix, I describe the flagellar family, protocols for purification of flagella, and microscopic techniques for the observation of flagella. Some lines might (unintentionally) resemble lines in "Flagella" chapters in the Encyclopedia of Microbiology and in the Encyclopedia of Genetics. It is not easy to write about flagella differently from previous writing.

I have to warn readers, again, that this analysis of the flagellar world is totally from my own perspective of the world. My knowledge about flagella was obtained mainly from *S. typhimurium*. Today, the *Salmonella* flagellum is the most extensively studied one, and I regard them as representative of eubacteria flagella in this book. I chose numerous papers on flagella and placed them in the References section, in which they were grouped according to each species. When the text in each chapter is too short to mention all papers involved or when there are not many papers to cite, the references are presented in order of the published years.

At the end of each chapter, I mention names of those who provided their strains to us, and the institutes they belong or belonged to (as indicated by "as of"). Their titles are omitted. Many of them were my good collaborators for years and then also gave me kind and encouraging comments for a draft of the chapter they were involved in for this book. I thank all the students who worked in my lab. Most of them were undergraduates and there are more than 100 altogether by now. I have been lucky to have so many students, which gave me an opportunity to think about as many projects as possible and thus turned an otherwise monotonous world into a colorful one. In case their work has never been published in some other form, I mention names of students who worked on the project. I offer a special thanks to Kaoru Uchida, my last Ph.D. student, who learned all the EM techniques I have used and took several pictures anew for this book. I also thank Tatsuya Yamasaki for his illustrations that have been inserted to fill the page.

"It is very important to have a patron who loves your work," facetiously said Hirokazu (Q-chan) Hotani—who was actually my patron—in his memoir. He organized a big project twice in his time and allowed me to join both. I learned a lot about science and life from Bob Macnab; one of his lessons was "Write papers every single year, but not just for Science or Nature." Thanks to this lesson, I could publish papers every single year, including a couple of Science papers (I'm sorry, Bob. Now I know what you meant).

I have been writing Haiku (the shortest style of poem) for the last 30 years. It may be out of character for me not to make one for this occasion.

振り返る勇気を得たり夏の果
(Furikaeru/yuuki wo etari/Natsu no hate)
Finally got a courage/to look back/in the end of the summer
August 2013, the hottest summer ever

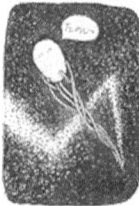   

Fun   Ooops   Ouch   No fun

(Depicted by Michiko Kobayashi, a mushroom painter)

1. BASIC KNOWLEDGE ABOUT FLAGELLA

The flagellum is an organelle of bacterial motility. It is a gigantic protein complex, consisting of three major substructures: the filament, the hook, and the basal body. The basal body includes an actively rotating part of the flagellum, the flagellar motor, which can generate torque from an electrochemical potential of proton gradient across the membrane, called proton motive force. Before going into chapters, I explain the basic knowledge and some unique terminologies only used for flagella, and introduce related Chapters and Topics in which you may find more references.

(A) Flagella Arrangement

Many (more than 70%) bacteria carry one or more flagella per cell and swim in water. Depending on the position of a flagellum growing on a cell, flagella are called with different names: polar (at one or both poles), subpolar (near the pole), lateral (from the middle half of the cell body), and peritrichous (randomly arranged on the cell body) flagella.[1]

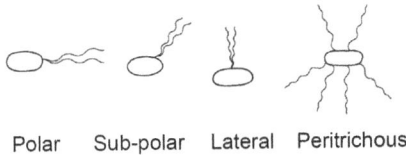

Polar   Sub-polar   Lateral   Peritrichous

*Figure I.1 **Flagella position.** (From left to right: Polar, Subpolar, Lateral, and Peritrichous flagella)*

In recent studies, peritrichous flagella and lateral flagella are often confused for historical reasons (see Topic: Flagella position and shape). Flagellar shape is not defined by its growing position, but is defined by the intrinsic properties of the flagellin (see Appendix: Flagellar family). Strictly speaking, the only example of lateral flagella included in this book is those of *Selenomonas ruminantium* (see Chapter 25). However, it is not easy to change history, and I, myself, sometimes use those names (pof for polar flagella and laf for lateral flagella) to distinguish two types of flagellins in one strain.

I do not use some more conventional names: monotrichous (single flagellum), multitrichous (more than two flagella), and lophotrichous (tuft at polar end of flagella) in this book. Instead, I use only the four positions mentioned above and the number of flagella: single, few, or numerous.

(B) Gram Staining

Gram staining of bacterial cells is neither an accurate nor elaborate technique, but nevertheless it is practically useful to distinguish two big domains of eubacteria species: Gram-positive bacteria that include most of the Firmicutes, and Gram-negative bacteria that include the rest. The structure of the flagellar basal body differs between the two types due to the difference of membrane structures; Gram-positives have two rings, while Gram-negatives have four rings (see next section). Accordingly, the genes encoding the two extra rings in Gram-negatives are missing in Gram-positives, as seen in *Bacillus subtilis* (see Chapter 4).

(C) *Salmonella enterica* Serovar Typhimurium

Salmonella enterica serovar Typhimurium had been just called *Salmonella typhimurium* until 2005 when the present name was proposed to classify members in a big *Salmonellae* family.[2] However, I will be talking about serovar Typhimurium only—and no other strains—among the *Salmonellae* family in this book. Therefore, hereafter, I will call this particular strain with the old shortened name, *Salmonella typhimurium*. When *S. typhimurium* and *E.coli* are dealt with to the same extent, I refer to them as *E.coli/Salmonella*.

2. FLAGELLAR STRUCTURE

(A) The Structure–Function–Gene Relationship in the Flagellum

The modern research on the bacterial flagellum started in 1974, when DePamphilis and Adler biochemically isolated the flagellar basal body and showed the electron microscopic images of the structure in a series of papers.[3–5] In 1985, I refined the purification method and established a protocol on how to purify the protein components of the basal body.[6] By the year 2000, most components of the flagellum had been identified, the pathway of flagellar assembly had been revealed, and roles of ca. 40 flagellar genes in the assembly process were now known.[7–9] The figure shows the structure–function–gene relationship in the flagellum.

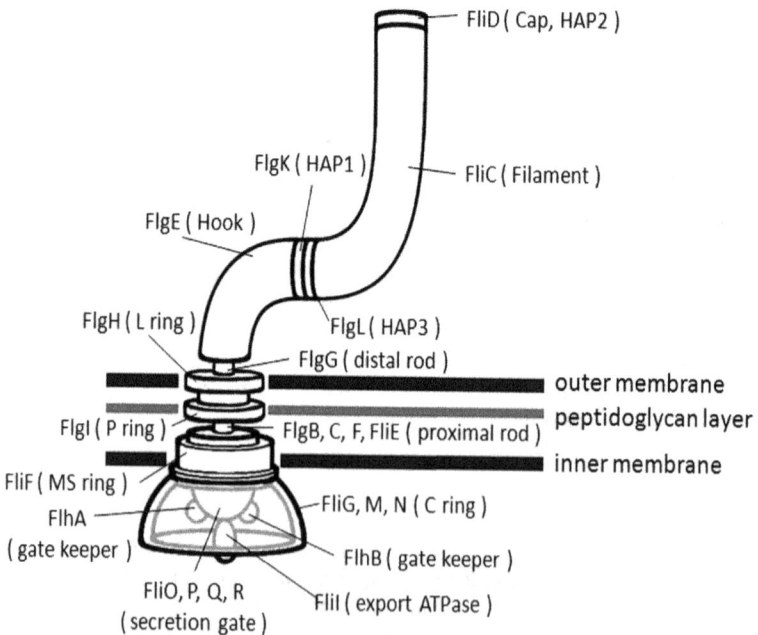

Figure I.2 **Structure of the flagellum.** *The name of a gene product is followed by the name of corresponding structure in parenthesis.*

The largest part of the flagellum is the filament, which is composed of thousands of subunits of one or multiple kind(s) of protein called flagellin or FliC. The reasons why a filament uses more than two flagellins are not completely understood (see Topic: Multiple flagellins). The filament is a helical tube. The question why one kind of chemically-identical protein can form a helix has not been fully answered and thus solving explains one of the mysteries about the flagellum (see Appendix: Flagellar family).

The filament is connected to the hook, which is hooked or sharply curved. Unlike the filament, the hook is always composed of a single kind of protein called hook protein or FlgE. The length of the hook is regulated to be short by secretion of a soluble protein called FliK (see Topic: Hook length). The mechanism of length control is still controversial.[10–13] FliK does not just regulate the hook length, but also indirectly regulates the gene expression of the late genes (*fliC, flgK, flgL*, and *che* genes) (see Topic: Gene regulation).

Between the filament and the hook, there is a small spacer called the HAP (hook-associated protein) region, which contains HAP1 (or FlgK) and HAP3 (FlgL). The HAP region is necessary for assembly of filaments and for stabilizing the filament shape as a helix.[14] Judging from the fact that HAP1 and HAP3 exist in any type of flagella, I suspect that the HAP region may play an important roles in conveying torque from the flexible hook (changing its conformation continuously during rotation) to the rigid filament.

The flagellum originates at the inner membrane and passes through the outer membrane or the cell wall. The structural entity for the anchoring in the membrane is called the basal structure or basal body. The basal body does not contain just those components necessary for motor function. In addition to the stator components (MotA and MotB), some fragile components have been detached from the basal body during purification. In 1985, one such fragile structure was found attached to basal bodies purified by a modified method; it was named the C (cytoplasmic) ring.[15,16] In 1990, another rod-like structure was found in the center of the C ring and named the C rod.[17] In 2006, flagellar export ATPase (FliI) was found at the periphery of the C ring as a complex with the supporter protein FliH.[18] Therefore, the basal structure (as of 2007) consists of the basal body, the C ring, the C rod, the export ATPase, and their regulators. There remain some genes with unknown functions even in *S. typhimurium*; they are *flhE, fliY* (not the same as *fliY* of *Bacillus* family, which is *fliM + fliN*), and *flk*.[19]

(B) Assembly Process of the Flagellum

The order of the steps toward the completion of a flagellum (the morphology pathway) has been analyzed in the same way as that which was used for bacteriophages: identifying intermediate structures in various flagellar mutants and aligning them in size from small to large ones.[20] Flagellar construction starts from the cytoplasm, progresses through the periplasmic space, and finally extends to the outside of the cell.

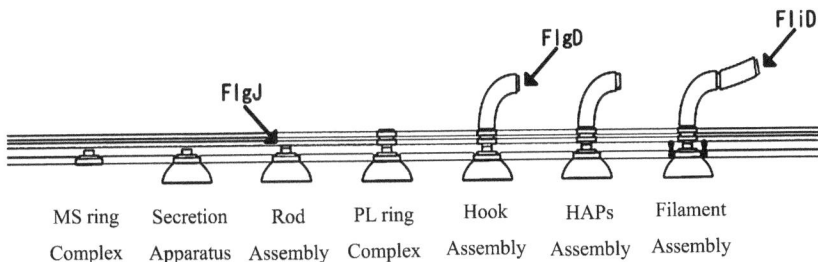

| MS ring | Secretion | Rod | PL ring | Hook | HAPs | Filament |
| Complex | Apparatus | Assembly | Complex | Assembly | Assembly | Assembly |

*Figure I.3 **Flagellar assembly process.** (Assembly proceeds from left to right, from small to large substructures.)*

In the Cytoplasm

In the assembly process, the very first flagellar structure recognizable by electron microscopy is the MS ring complex[21]; other components assemble on the MS ring complex one by one. Therefore, the MS ring complex is regarded as the construction base for the flagellum. When two other flagellar substructures, the C ring and the C rod, are constructed on the cytoplasmic side of the M ring, the gigantic complex starts secreting other flagellar proteins to construct the extracellular structures of the flagellum.

In the Periplasmic Space

The first extracellular structures constructed on the MS ring complex are the proximal rod and then the distal rod. When the rod has grown large enough to reach the outer membrane, the hook starts growing. However, the outer membrane physically hampers the hook growth until the outer ring complex makes a hole in it. Among flagellar proteins, FlgH and FlgI, the component proteins of the outer ring complex, are exceptional in terms of the manner of secretion: these two proteins have cleavable signal peptides and are exported through the general secretion pathway.[22] However, under special conditions, filaments grow in the absence of the outer rings but stay in the periplasm to form the periplasmic flagella as seen in spirochetes (see Chapter 6).

Outside the Cell

Once the physical block by the outer membrane has been removed by the outer PL rings, the hook resumes growth with the aid of FlgD until the length reaches ca. 55nm (see Topic: Hook length). Then, FlgD is replaced by HAPs, which is followed by the filament growth. The filament growth proceeds only in the presence of FliD (HAP2 or filament cap protein); without this cap, exported flagellin molecules are lost to the medium.

In conclusion, a flagellum grows from bottom to tip.[23] The component proteins of the axial structures (the rod, hook, and filament) do not retain the signal peptides, but are secreted without cleavage through a special secretion system of the flagellum. Today it is called the type III secretion system (T3SS), which was originally used for one type of virulence secretion system in invasive pathogenic species and adopted for the flagellar secretion system due to the structural and functional similarities between the two.[24] The axial structures do not self-assemble *in vivo*, but require helper proteins or cap proteins: FlgJ for the rod, FlgD for the hook, and FliD for the filament (find them in the Figure above). The other type of helper proteins are chaperones: FlgA is a chaperone for P-ring formation, FliS for flagellin, FliT for FliD, and FlgN for FlgK and FlgL. Flagellar assembly is not a simple aggregation of proteins, but a sophisticated assembly system with self-controlled devices far more developed than any other cell-surface structures.

3. FLAGELLAR GENETICS

(A) Unified Gene Names for *E. coli/Salmonella*

Names of the flagellar genes were originally specific for each of these two species. But when the number of genes identified year after year increased over the number of alphabets, it was necessary to rename the genes. By then, people noticed the similarity of gene function between *E. coli* and *S. typhimurium*. Bob Macnab proposed that flagellar geneticists use unified nomenclature for the flagellar genes for both species. After tough negotiation, the proposal was successfully accepted.[25] I summarize the unified genes together with old names, and their functions on a list, which will serve as a good guide when you have to enter the forest of the classic papers written before 1985.

Table I.1 Unified Flagellar Gene Names for *E. coli* and *S. typhimurium*

Old Symbols		New Symbols	Structure or Function
E. coli	*S. typhimurium*		
flaU	*flaFI*	*flgA*	Periplasmic chaperone for P-ring formation
flbA	*flaFII*	*flgB*	Proximal rod protein (Rod 1)
flaW	*flaFIII*	*flgC*	Proximal rod protein (Rod 1)
flaV	*flaFIV*	*flgD*	Cap protein for hook growth
flaK	*flaFV*	*flgE*	Hook protein
flaX	*flaFVI*	*flgF*	Proximal rod protein (Rod 1)
flaL	*flaFVII*	*flgG*	Distal rod protein (Rod 2)
flaY	*flaFVIII*	*flgH*	L-ring protein
flaM	*flaFIX*	*flgI*	P-ring protein
flaZ	*flaFX*	*flgJ*	Cap protein for rod growth, muramidase
flaS	*flaW*	*flgK*	HAP 1
flaT	*flaU*	*flgL*	HAP 3
*	*	*flgM*	Anti-sigma factor
*	*	*flgN*	Chaperone for FlgK, FlgN
flaH	*flaC*	*flhA*	Secretion gate keeper
flaG	*fiaM*	*flhB*	Secretion gate keeper
flaI	*flaE*	*flhC++*	Regulation of gene expression
flbB	*flaK*	*flhD++*	Regulation of gene expression
*	*	*flhE++*	Required for swarming motility
flaD	*flaL*	*fliA*	Sigma factor 28
–	*nml*	*fliB++*	N-methylation of lysine residues in flagellin
hag	*HI*	*fliC*	Flagellin
flbC	*flaV*	*fliD*	HAP 2 or cap protein for filament growth
flaN	*flaAI*	*fliE*	Proximal rod protein (Rod 1)
flaBI	*flaAII.1*	*fliF*	MS-ring protein
flaBII	*flaAII.2*	*fliG*	C ring, torque generation by interacting with MotA
flaBIII	*flaAII.3*	*fliH*	Anchoring FliI to FliN
flaC	*flaAIII*	*fliI*	ATPase for T3SS
flaO	*flaS*	*fliJ*	Interact with FlhA
flaE	*flaR*	*fliK*	Hook length control, switching of secretion substrate specificity
flaAI	*flaQI*	*fliL*	Basal body-associated membrane protein
flaAII	*flaQII*	*fliM*	C ring. Motor switching by interacting with CheY-Phosphate
motD	*flaN*	*fliN*	C ring
flbD	*flaP*	*fliO++*	FliP integrity
flaR	*flaB*	*fliP*	Cytoplasmic (C) rod, Secretion gate

(Continued)

Old Symbols		New Symbols	Structure or Function
E. coli	*S. typhimurium*		
flaQ	flaD	fliQ	C rod, secretion gate
flaP	flaX	fliR	C rod, secretion gate
*	*	fliS	Chaperone for FliC
*	*	fliT++	Chaperone for FliD, binding to FlhDC to suppress
*	*	fliY++	Gene activator
*	*	fliZ++	Binding HilD to repress SPI1 genes
*	rflH	flk++	Secretion of FlgM and hook length control

Table I.1 (Continued)

*These genes were identified after unification of gene names.
++These genes are not general, but are specific for Salmonella and related species.

I have to admit that I did not pay much attention to chemotaxis in this book. Please see a few recent review articles to supplement this work with that important half of flagella research.[26–28]

(B) Differences between *E. coli* and *S. typhimurium*

The number and arrangement of flagellar genes are almost identical in both *E.coli* and *S. typhimurium*. However, *S. typhimurium* has five more genes than *E.coli*: *fliB, fljA, fljB, hin,* and *flk*. FliB is methyltransferase for *S. typhimurium* flagellin and the *E. coli* flagellin is not methylated. The *fljA, fljB, hin* are used for phase variation. *Salmonella* species have two sets of flagellin genes, *fliC* and *fljB*. But, we do not count these flagellins among the multiple flagellins (see Topic: Multiple flagellins), because they are not incorporated into the same filament. The *hin* gene—upstream of the *fljB* gene—flip-flops, allowing *fljB* gene expression in only one direction. When the *fljB* gene is expressed, the *fljA* gene downstream of the *fljB* produces a repressor of the *fliC* gene, inhibiting a concomitant expression of the latter. By switching these flagellins, *Salmonella* cells can produce two kinds of flagella with two different antigens and evade the immune system of the host. This phenomenon is called phase variation.[29,30] The function of the *flk* gene is involved in hook length control, but the details are still ambiguous.[31,32]

(C) Extra Genes: Species-Specific Flagellar Genes

There are several genes that are not included in the unified nomenclature of *E.coli/Salmonella*, but exist in other species and have their own names, such as *fla, flb, fle, flr,* and *pom*. There are several patterns for calling new genes of these new names:

1. They could be truly new flagellar genes that do not exist in *E.coli/Salmonella* species.
2. They may have a similar function to *E.coli/Salmonella* genes, but are not quite the same.
3. They are just mistakenly annotated.

I closely examined those extra genes in each species and calculated the homology with the preexisting genes. Many are left as they were, but some of them are annotated to conventional genes (see below: the *fliC* genes).

Table I.2 Extra Genes

Gene names	Other names	Function/Structure	Oorganism
cheL		Chemotactic signal response	B. japonicum, B. oligotrophica, G. oxydans, M. magnetotacticum, R. sphaeroides
cheV		cheA activity modulation	X. oryzae
cheX	~cheC	phosphatase	B. burgdorferi
flaA	fliC	flagellin	A. brasilense, A. fischeri, H. pylori, R. sphaeroides, Ruegeria sp.
flaE		flagellin	B. japonicum, C. crescentus, A. fischeri
flaF		flagellin	B. japonicum, A. brasilense, B. oligotrophica, C. crescentus, R. sphaeroides, G. oxydans, Rugeria sp., S. thermophilum
flaG		In a unit: fliC–flaG–fliSD	A. fischeri, B. japonicum, B. subtilis, C. crescentus, E. casseliflavus, G. kaustophilus, H. pylori, I. loichiensis, Paenibacillus sp., P. aeruginosa, S. ruminantium, S. degradans, V. parahaemolyticus
flaY		unknown	C. crescentus
flbA	flhA	unknown	B. burgdorferi, C. crescentus, G. oxydans
flbB	flhD	Transcriptional activator	B. burgdorferi
flbD		Transcriptional activator	A. missouriensis, A. brasilense, B. burgdorferi, B. bacteriovorus, C. crescentus, E. casseliflavus, G. kaustophilus, Paenibacillus sp., S. ruminantium
flbT		Repressor, post-trascriptional	A. brasilence, B. japonicum, B. oligotrophica, C. crescentus, M. magnetotacticum, R. sphaeroides, Rugeria sp.
flbY		HBB component	B. japonicum, C. crescentus
fleE		unknown	G. oxydans
fleN	flhG	Multiflagella	P. aeruginosa
fleQ		Master regulon	L. pneumophila, P. aeruginosa, R. sphaeroides, S. degradans, X. oryzae
fleP		Long pili	P. aeruginosa
fleR		Class III regulator	L. pneumophila, P. aeruginosa
fleS		Class III regulator	P. aeruginosa
flgTOP		unknown	V. parahaemolyticus
flhF		a signal recognition particle -type GTPase	H. pylori, Paenibacillus sp., S. ruminantium, R. sphaeroides, B. subtilis
flhG	fleN	ATPase, regulation of the number of flagella	B. bacteriovorus, H. pylori, P. aeruginosa, Paenibacillus sp., S. ruminantium
fliW		Chaperone / Antagonize CsrA	A. missouriensis, A. brasilense, B. burgdorferi, B. subtilis, B. bacteriovorus, G. kaustophilus, H. pylori, Paenibacillus sp., S. ruminantium
fliX		In a unit: fliX–flgI	A. brasilense, B. japonicum, B. oligotrophica, C. crescentus, M. magnetotacticum

(Continued)

Table I.2 (Continued)

Gene names	Other names	Function/Structure	Oorganism
flrA		σ54-dependent activator	A. fischeri
flrB		Sensory kinase	A. fischeri , L. pneumophila
flrC			A. fischeri
fgtA		glycosylation	P. aeruginosa
pflA	flhF		H. pylori
pflI			C. crescentus
lafK			V. parahaemolyticus
lafKWZ			E. coli
motC	cheV		P. aeruginosa
motD			P. aeruginosa
motE		Chaperone for motC	S. meliloti
motS		Na-driven motor	B. subtilis
motP		Na-driven motor	B. subtilis
pomA		Na-driven motor in polar flagella	B. japonicum
pomB		Na-driven motor in polar flagella	B. japonicum
flaAB		flagellin	H. pylori
flaABCDEF		flagellin	V. parahaemolyticus
flaMLKJ			V. parahaemolyticus

(D) Functional units: Clusters of Functionally Related Genes

Several genes that are functionally related to each other often appear as a small cluster in the genome. These groups of genes are found in all species, suggesting that a small block of gene products are required for assembly of flagellar substructures and that those functional units might have evolved together.

Table I.3 Functional Units

Gene Unit	Function	Role of each Gene
flgAHI	LP ring formation	FlgA (chaperone for P ring), FlaH (L ring), FlgI (P ring)
flgBC	Rod1 formation	FlgB (rod 1 protein), FlgC (rod 1 protein)
flgFG	Rod2 formation	FlgF (rod 1), FlgG (rod 2)
flgDE fliK	Hook formation	FlgD (hook cap protein), FlgE (hook protein), FliK (hook length control)
flgKL	Hook—filament junction (HAP region)	FlgK (HAP1), FlgL (HAP3)
flhAB	Secretion gate-keeper	FlhAB (secretion gate-keeper protein)
flhFG	Control of number and position of polar/sub-polar flagella	FlhF (a signal recognition particle -type GTPase: polar location determinant), FlhG (a putative ATPase, binding to FlhF to prevent its role)
fliCDS	Filament formation	FliC (flagellin), FliD (HAP2/ filament cap protein), FliS (chaperone for flagellin)

(Continued)

Table I.3 (Continued)

Gene Unit	Function	Role of each Gene
fliFG	MS/-C ring formation	FliF (MS ring), FliG (a C ring component)
fliHIJ	Transport complex	FliH (Anchoring FliI to FliN), FliI (ATPase), FliJ (Interact with FlhA)
fliMN(Y)	C ring formation	FliMN (C ring component), FliY (FliM + N)
fliPQR	Secretion gate	FliPQR (secretion gate protein)
Joint units often observed		
flgFGAH	Rod-L ring formation	FlgA move together with FlgH more than with FlgI.
flhABFG	Gate-keeper and positioning of flagella	Although there is no obvious functional relationship between FlhAB and FlhFG, they often move together in the genome.
fliC (flaG) fliDS	filament formation	The function of FlaG is unknown.

Using the functional units, organization of flagellar genes can be analyzed in types of structural units as follows:

[*E. coli/Salmonella* type]

Region I/ *flg*: (rod1) (hook) (rod2) (PL ring) (HAP region),
Region II/ *flh*: (secretion gate keeper) (chemosensor) (motor stator) (master genes),
Region III/ *fli*: (filament) (rod1) (MS-C ring) (transport ATPase complex) (export regulation) (C ring) (secretion gate)

[*Bacillus* type]

The first cluster: (rod1) (MS-C ring) (transport ATPase complex) (hook) (C ring) (secretion gate) (secretion gate keeper) (chemosensor)
The second cluster: (filament) (HAP region)

[*Helicobacter* type]

Although the gene arrangement is the most random in this organism, several functional units are still kept in the genome; they are (MS-C ring) (filament cap) (rod1) (hook) (see Topic: Gene arrangement).

4. EM TECHNIQUES FOR FLAGELLA STUDY

(A) PTA Staining

The images of bacterial cells and flagella in this book were obtained with the transmission electron microscope (JEM-1200EXII, JEOL, Tokyo). Flagellar filaments are very thin as compared with the cell body. To clearly visualize filaments on a cell, cells are washed to remove salts and/or detergents in the sample just before staining. Samples were negatively stained with a 1% (for a whole cell) or 2% (for flagella only) solution of phosphotungstic acid (PTA) adjusted at pH 7.0. For the negative staining, 0.5% uranyl acetate solution (UA) has long been used, but today the usage is limited in Japan due to the radioactivity. There are merits and drawbacks to each staining method. UA gives a stronger contrast of images than PTA. The UA solution is acidic (pH 4), however, while the pH of PTA is adjustable to any value from acidic to alkaline. Flagellar filaments are sensitive to acidic pHs; the filament shape transforms from one to another according to the pH values, and most of them start to depolymerize at around pH 4. How the filament structure is damaged by UA is shown in the following Figure.

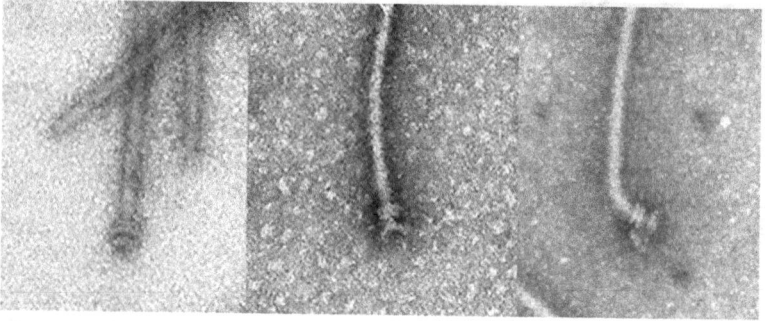

*Figure I.4 **The pH effects of the staining solution on flagellar filaments.** (From left to right: UA (pH 4.0), PTA (pH 4.0), and PTA (pH 7.0))*

Although the contrast of images stained with PTA is weaker than those with UA, substructures of flagellum (hook, HAP region, and filament) can be clearly distinguishable by PTA staining.

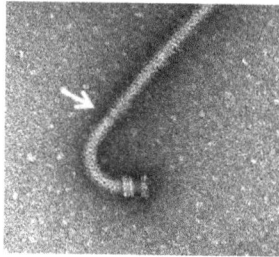

*Figure I.5 **Basal structure stained by PTA at pH 7.** (Arrow indicates HAP region between the hook and filament.)*

More importantly, we observe that cells in PTA solution are as active as they are in water. We do not fix samples with glutaraldehyde or similar reagents. We first put a drop of PTA solution on an EM grid and then add a tiny amount of cell suspension into the drop and let it sit for 1 min. Consequently, *S. typhimurium* cells, for example, are actively swimming in a PTA solution at neutral pH as confirmed by optical microscopy. When the solution on the grid is sucked up with a piece of the filter paper, most cells spread flagellar bundles into an open position on an EM grid. However, if lucky, you will find cells that keep bundles behind the cell body as if they were still swimming. Altogether, in order to observe intact flagella, I personally prefer PTA to UA.

*Figure I.6 **A drop of staining solution on a grid.** (On the grid at the tip of a forceps, we put a drop of PTA solution and add a drop of cell suspension. In the PTA solution, cells are still swimming.) **EM picture of a "swimming" cell.** (A cell having a flagellar bundle behind the cell)*

(B) Osmotically-Shocked Cell Method

In order to visualize the flagellar basal body buried in the cell wall or the membranes, we often employ osmotically-shocked cells, in which the cell contents are eluted out and the cells appear transparent (See Appendix: protocol IV for preparation of osmotically-shocked cells). Here, I show one example of an osmotically-shocked cell prepared by protocol IV. In the picture, you see numerous SPI1 needle complexes, *Salmonella* virulence secretion apparatus (see Chapter 24), which are too short to observe on the cell surface just stained by the ordinary method.

Question: How many needle complexes can you find on the cell surface and in the cell (actually, they are on the cell surface of both this side and the other side of the cell)?

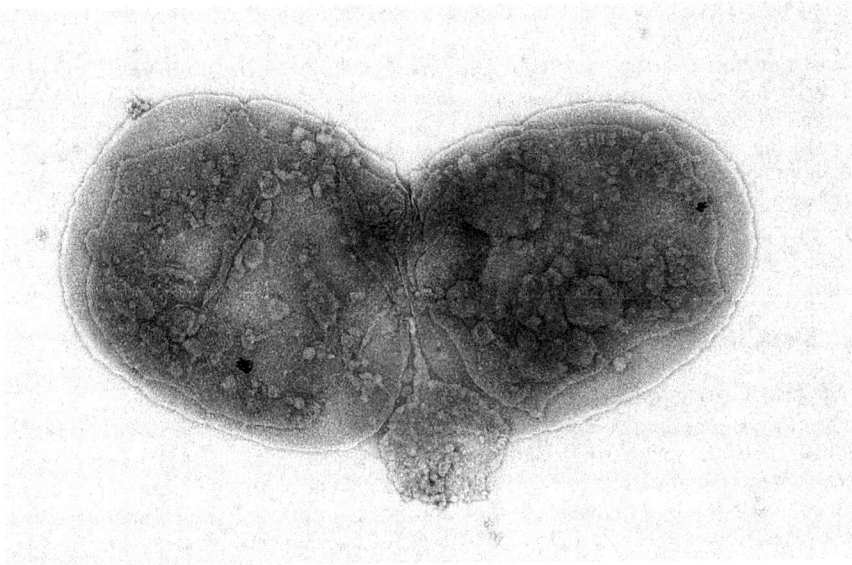

Figure I.7 **EM picture of an osmotically-shocked cell.** (*A* Salmonella *cell dividing into half.*)

Answer: I count at least fourteen needle complexes on the cell surface and seven in the cell.

CHAPTER EXAMPLES

1. OUTLINE

Each chapter contains data for one species in two facing pages except for several species such as *E. coli* and *S. typhimurim* (four pages for each). On the first page, a photograph of a whole cell of the wild-type strain is shown. The scale bar is typically 1 µm. The phylum is shown on the right of the cell photograph. In the bottom half, pictures of flagella on the cell and/or isolated flagella are shown. In case the strain holds other cell-surface appendages, those photograph are also shown. I did not dare to put scale bars in each photograph to avoid complexity.

On the second page, the genetic map regarding flagellar genes is shown. In making even numbers of pages for each chapter, I put several Topic pages to fill the gap of pages. Many of the topics are a summary of the comparison of physical properties of a flagellar component among species. What is the largest flagellin? Which has the longest hooks? and so on.

2. SPECIFIC INDICATIONS

(A) The Genetic Map

When the genome data on a strain we analyzed is incomplete or not open to the general public, I used the genome information of closely related species. Flagellar genes are collected from several databases (KEGG, NCBI, and PDB), and the accession numbers are given. Under the strain name, total number of nucleotides and total number of protein genes are cited. This is a physical map of the flagellar genes, that is, the gene position is indicated by the nucleic numbers (kbp) rather than by the accession numbers.

(B) The Flagellin Gene (*fliC*)

The flagellin(s) is/are shown in bold, because a flagellin has several names different from FliC. Sometimes I changed annotation of certain genes in a databank into FliC by calculating sequence homology and comparing the arrangement of neighboring genes. Some species have multiple flagellin genes with special names other than *fliC*. If those flagellins were experimentally confirmed to exist in the flagellum, I left their original names as they are; for example, *fljJ-fljO* in *C. crescentus* (see Chapter 8), and *flaA/flaB* in *B. burgdorferi* (Chapter 6) and *H. pylori* (Chapter 13). These gene names are not included on the list of Extra genes. Otherwise, I automatically named them *fliC1*, *fliC2*, and so on.

(C) Duplication of Genes

Some genes appear more than twice in a genome. Genes with homologous sequences have subscript numbers. When a duplicated gene was truncated, the gene is marked with the prime symbol: *fliG'*. But many of them are left as they are without comments.

(D) The *che* Genes

There are many *che*-related genes found in genomes of all species[1]; *mcp* (Methy-accepting-protein) are especially abundant. Some of the *che* genes are probably not for the flagella system, but for other two-component systems. I cited only those located among the flagellar gene clusters or those that are experimentally confirmed.

(E) Missing Genes

There are genes essential for assembly of the flagellum.[2] They encode proteins that form substructures of the flagellum: C ring, MS ring, rod, hook, HAP region, and filament. All flagella so far studied are composed of these substructures. On the other hand, regulatory genes are found in some species, but not in the others; for example, the *flhCD* genes are the master genes that control the expression of the other flagellar genes in *E. coli/Salmonella* and related species. Chaperones *fliT* and *flgN* are not essential.[3,4] Gram-positives do not have the PL ring complex, thus *flgA, H, I* are essential for Gram-negatives only. *fliB, fliO, fliT, fliY,* and *fliZ* are specific for *S. typhimurim* and some related species.[5] In summary, I regard the following genes the essential genes for functional flagellum. Species-specific genes are indicated in parentheses.

> *flg(A),B,C,D,E,F,G,(H)(I),J,K,L,M,(N),flhA,B,(flhF,G),*
> *fliA,(B),C,D,E,F,G,H,I,J,K,L,M,N(or Y),(O),P,Q,R,S,(T),(Y),(Z) motA,B.*

In the Missing genes section in each chapter, I picked only those genes that are essential, but missing in the genome map. Most of them are genes for flagellar substructures as confirmed by EM and thus should be found in the future. The *flhFG* genes are required for placing flagellum at the pole.[6] When we observe cells with polar or subpolar flagella, we expect the *flhFG* genes to exist somewhere in the genome.

(F) Extra Genes

Some genes that are not found in *E. coli/Salmonella* are marked with parentheses. Those extra genes are summarized on a list in the Introduction.

(G) Direction of Transcription

Operons and the direction of transcription of the flagellar genes have been experimentally determined in *E. coli/Salmonella*.[7] Therefore, arrows indicating the direction of transcription are shown only in the genomes of *S. typhimurim* (Chapter 24) and *E. coli* (Chapter 9). In other species, the direction of transcription should be read according to the order of the alphabet as in *S. typhimurim*; for example, *fliFGHIJ* is on the (+) strand, whereas *fliJIHGF* is on the (−) strand.

CHAPTER 1

Actinoplanes missouriensis — Swimming Spores with Flagella

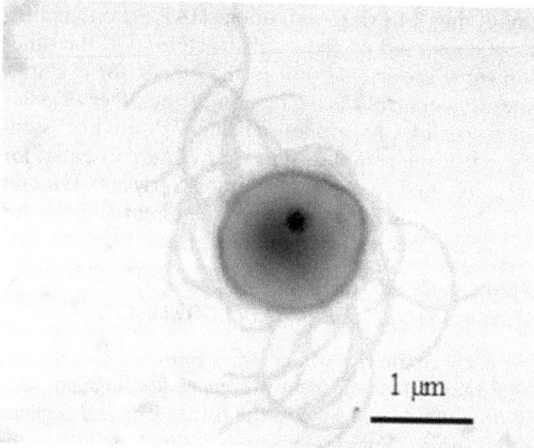

1 μm

Phylum	Actinobacteria
Class	Actinobacteridae
Order	Actinomycetales
Family	Micromonosporaceae
Genus	Actinoplanes
Species	*A. missouriensis*

Actinoplanes missouriensis is a Gram-positive, soil-inhabiting, filamentous bacterium that characteristically produces spores within a terminal sporangium. Spores are released from sporangia upon contact with water (containing soil extracts) and actively swim by means of flagella, which are completed when the spore is matured. The swimming speed of an *A. missouriensis* cell is as high as 135 μm/s. Compare this speed with that of the fast swimmer in the sea (Chapter 22).

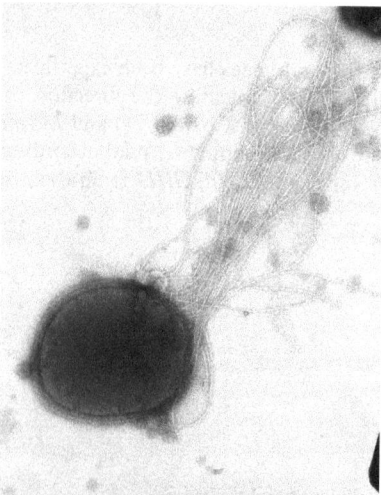

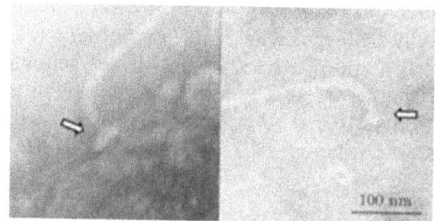

100 nm

The basal body is composed of MS rings, but is missing the outer PL rings: a common feature for Gram-positive Firmicutes. The filament is composed of a single kind of flagellin with the MW of 44 kDa.

A. missouriensis flagella grow from one side of the spore rather than from all over, indicating that the spore flagella are not peritrichous, but either polar or lateral despite its round cell body (see Topic: Flagellar position and shape). This kind of flagella arrangement also appears in *M. voltae* (see Chapter 32).

Actinoplanes missouriensis 431 genome **(8,773,466 bp/ 8,124 genes)**	NC_017093 AP012319 AB583923

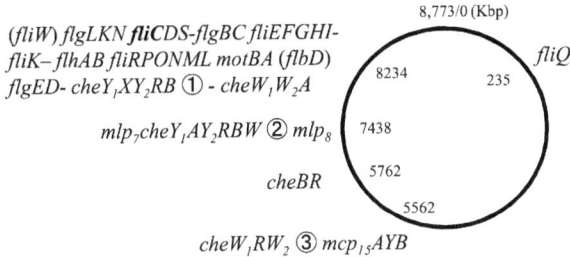

(*fliW*) *flgLKN* **fliCDS**-*flgBC fliEFGHI-fliK–flhAB fliRPONML motBA* (*flbD*) *flgED*- *cheY₁XY₂RB* ① - *cheW₁W₂A*

mlp₇cheY₁AY₂RBW ② *mlp₈*

cheBR

cheW₁RW₂ ③ *mcp₁₅AYB*

8,773/0 (Kbp)

8234   235   *fliQ*

7438

5762

5562

Missing genes: *flgF,G,M, fliA,J*

The genome of *A. missouriensis* 431 strain consists of a single 8.7-Mbp chromosome. Flagellar genes form a compact cluster at 8,234 bp on the chromosome. Almost all flagella structural genes except those for the rod proteins are found in the compact cluster. There are 3 complete sets of *che* genes. More than 20 *mcp*, *mlp* (*mcp*-like-protein) genes are scattered all over the chromosome.

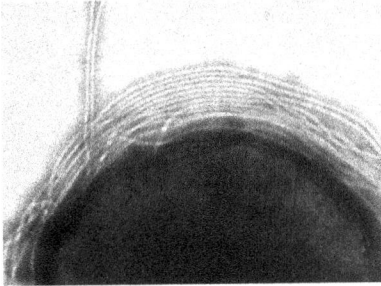

The average number of flagella per spore is 15, and the average length of filaments is 1.89 µm, which covers the circumference of the cell body. Release of spores from sporangia starts upon contact with water. 10 minutes after contact, numerous spores had flagella which looked to be forming several layers around the spore. *Actinoplanes* spores eventually grow into mycelia and form sporangia again under adverse conditions.

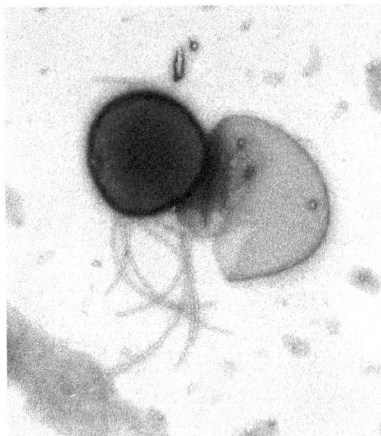

Two types of spores were observed in the preparation of PTA-negatively-stained samples: one appears darkly stained, and the other is lightly stained (lower). The latter appears to be a membranous empty sheath with a large crevice in the middle of the body, indicating that flagellar assembly has already been completed in the sheath. Flagellated spores hatch and leave an empty sheath behind.

Strains were provided by Yasuo Ohnishi of the University of Tokyo, Tokyo, Japan. The project was carried out by Moon Sun Jang and Kaoru Uchida.

CHAPTER 2

Aliivibrio fischeri — Light-Organ Symbiont in the Bobtail Squid

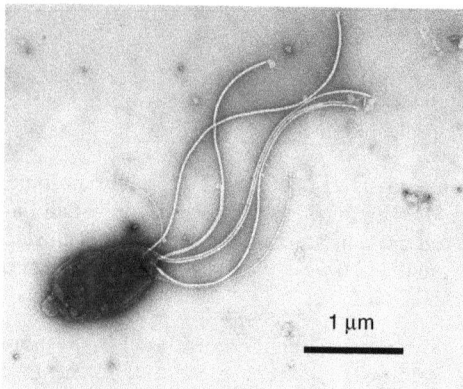

Phylum	Proteobacteria
Class	Gammaproteobacteria
Order	Vibrionales
Family	Vibrionaceae
Genus	Aliivibrio
Species	*A. fischeri*

1 µm

Aliivibrio (formerly *Vibrio*) *fischeri* is a Gram-negative, marine bacterium. *A. fischeri* has a symbiotic relationship with the bobtail squid (*Euprymna scolopes*), in which it produces **bioluminescence** from the host's luminous organ. The light emission is controlled by the bacterium's *lux* operon. Each cell possesses **several polar flagella**. *A. fischeri* has only one flagella system, whereas other *Vibrio* species often have two flagella systems: polar flagellum and lateral flagella (see Chapter 28. *V. parahaemolyticus*).

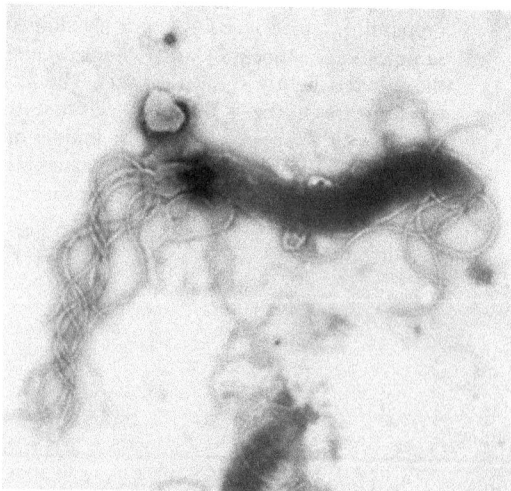

The polar flagella of *A. fischeri* are sheathed. **The filament is composed of six flagellins (FlaA–F)**, which have similar amino acid sequences (~65 % homology). By analysis of deletion mutants of each gene, four of the flagellins have been found in a filament: all mutants produce flagella and swim. A *flaD*-deletion mutant (left) produces numerous polar flagella with a smaller pitch than that of wild-type.

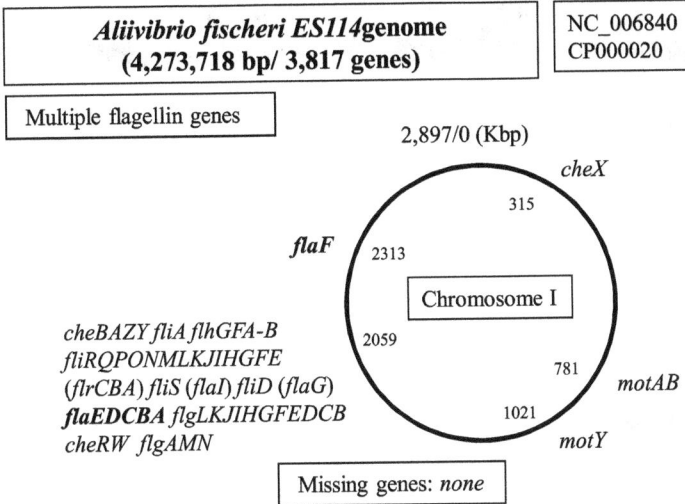

| **Aliivibrio fischeri ES114 genome** (4,273,718 bp/ 3,817 genes) | NC_006840 CP000020 |

Multiple flagellin genes

2,897/0 (Kbp)

cheX

315

flaF   2313

Chromosome I

2059

781   *motAB*

1021

motY

*cheBAZY fliA flhGFA-B fliRQPONMLKJIHGFE (flrCBA) fliS (flaI) fliD (flaG) **flaEDCBA** flgLKJIHGFEDCB cheRW flgAMN*

Missing genes: *none*

The genome of *V. fischeri* ES114 strain consists of two chromosomes (I: 2,897,536 bp and II: 1,330,333 bp) and one circular plasmid (45,849 bp). Flagellar genes form one of **the most compact clusters** in a locus at 2,059 Kbp on chromosome I (see Topic: Gene arrangement). Flagellin genes (*flaA–E*) are adjacent to each other, but *flaF* is located alone away from the flagellar cluster. There is another set of *motAB* genes on chromosome II (not shown). The role of a single *motY* at 1,021 Kbp is not known. The *flrA* is a sigma 54-dependent activator, and *flrB* is a sensory kinase (see *L. pneumophila*). Most of the *lux* genes required for bioluminescence are clustered on chromosome II.

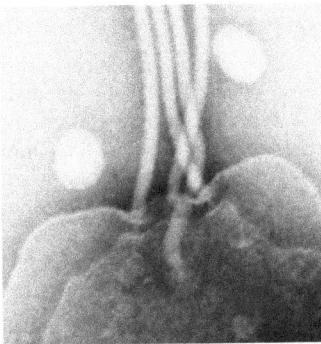

The flagellar genes of *V. fischeri* ES114 strain are more compactly clustered in one locus than are those of *Salmonella*. The basal structure (left) appears similar to that of *S. typhimurium* by electron microscopy.

Strains were provided in 2005 by Ned Ruby of the University of Hawaii (present address is University of Wisconsin–Madison), USA.
The project was carried out by Shutaro Ochikubo and Rika Tasaki.

CHAPTER 3

Azospirillum brasilense — A Bushy Hook of the Polar Flagellum

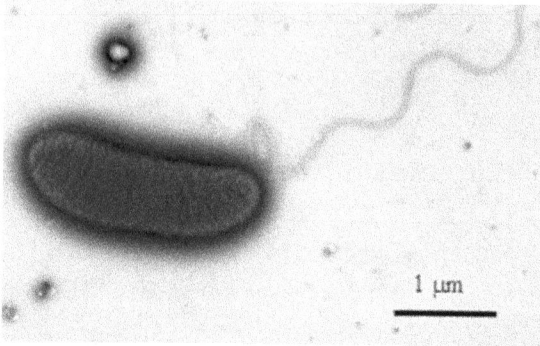

Phylum	Proteobacteria
Class	Alphaproteobacteria
Order	Rhodospirillales
Family	Rhodospirillaceae
Genus	Azospirillum
Species	*A. brasilense*

Azospirillum brasilense is a Gram-negative, nitrogen-fixing bacterium that colonizes the rhizosphere of various agronomically important grasses and cereals. This strain has **one of the largest genomes in the phylum Proteobacteria** and possesses two sets of flagellar systems: **a polar flagellum and lateral flagella**, in a similar manner to *Vibrio spp*. The dominant motility response of *A. brasilense* is a metabolism-dependent form of taxis known as **energy taxis**.

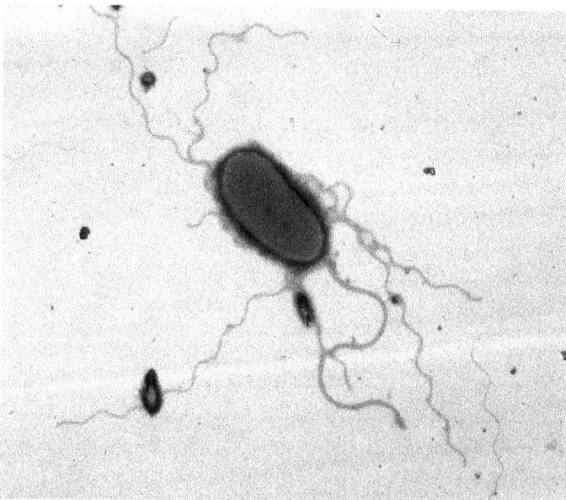

A cell produces two types of flagella depending upon growth conditions: a polar flagellum (thick arrow) alone when grown in a liquid medium (above), and both a polar flagellum and lateral flagella (thin arrow) when grown in a viscous medium (left). It should be noted that the helical pitch and diameter of the polar flagellum is much larger than those of the lateral flagella (see Appendix: Flagellar family).

**Azospirillum brasilense Sp245 genome
(7,530,241bp/ 7,557 genes)**

Multiple flagellin genes

NC_016617
HE577327

3,023/0(Kbp)

fliQ₁

618

798   *fliG*

cheY₃AWmcp cheRDB   2993

*flhB fliRQ₂E flgCB-flhB'--fliP--
motB₁ cheZ-(flaA)fliML-
flgFGAH---(fliX)flgIJ fliK flgD*

Chromosome

2341

1607

NC_016595
HE577330   778/0 (Kbp)   *fliN*

The polar flagellum

NC_016594
HE577328

flgE₂E₃flgD   573   Plasmid 3

fliSD fliC2 flgLK   567   1,766/0(Kbp)

fliC1 (flbT)   1624

flhGFA (flbD) motA fliNHF   1283

fliI   1277   Plasmid 1

cheA₂WYB₂R   1272

1259   523   *flgE₁K*

Missing genes: *flgM, fliA,J*   *motB₂*

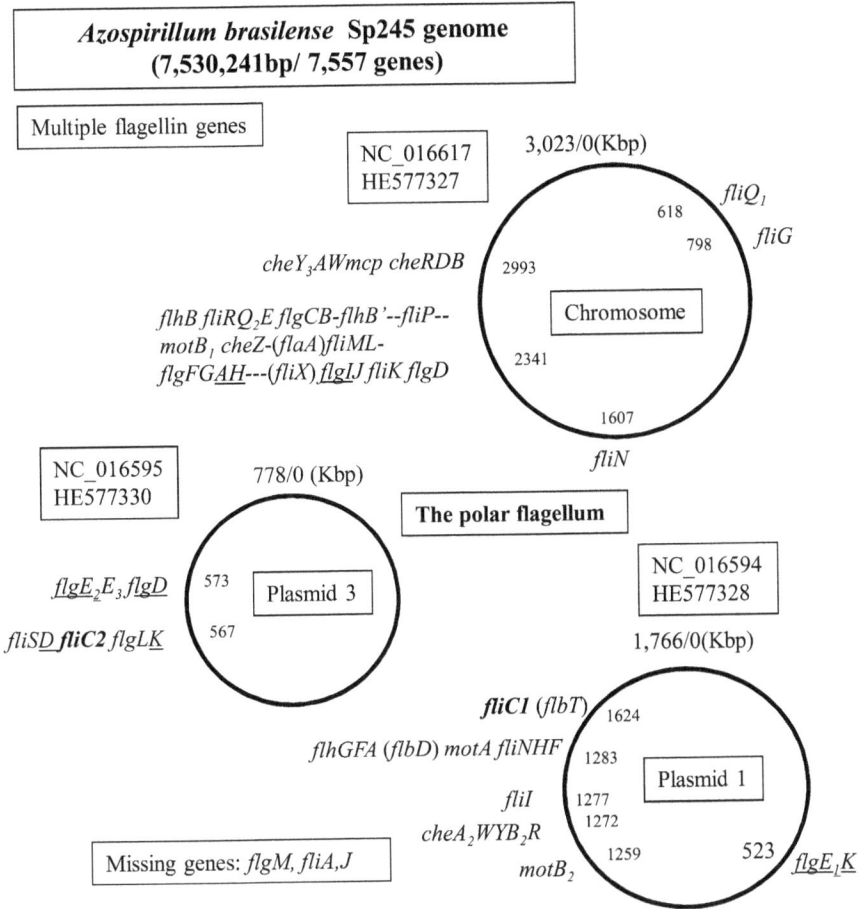

The genome of *A. brasilense* Sp245 strain consists of one 3,023,440 bp chromosome and six large circular plasmids. Total genome size is 7,530,241 bp. Genes necessary for assembly of **the polar flagellum** form a big cluster on the chromosome and small clusters at several scattered loci on plasmid 1 and 3 (above), while those for the lateral flagella are scattered at several loci on plasmid 4 (next page). Arrangement of flagellar genes is irregular and random, much as in *B. japonicum* (see Chapter 7). There are **three flagellin genes**: *fliC1* on plasmid 1, *fliC2* on plasmid 3, and *laf* (417aa) on plasmid 4. There are three Mot proteins for the polar flagellum: one MotB₁ on the chromosome, and one MotA and MotB₂ on plasmid 1.

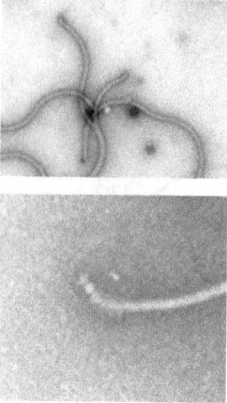

Purified **polar flagella** give rise to a single band at ~95 kDa by SDS-PAGE. **The flagellin is heavily glycosylated**, as revealed by periodic acid-Schiff staining. Judging from the molecular size and the location of the gene, the polar flagellin is FliC2 (621aa), agreeing with an observation that one of the unglycosylated flagellin is 65 kDa. Whether FliC1 (274aa) is incorporated into the polar flagellum is not known.

The purified polar flagellum (lower panel). The most striking feature of the polar flagellum is the hook; by electron microscopy, **the hook appears to be covered with bushy materials**.

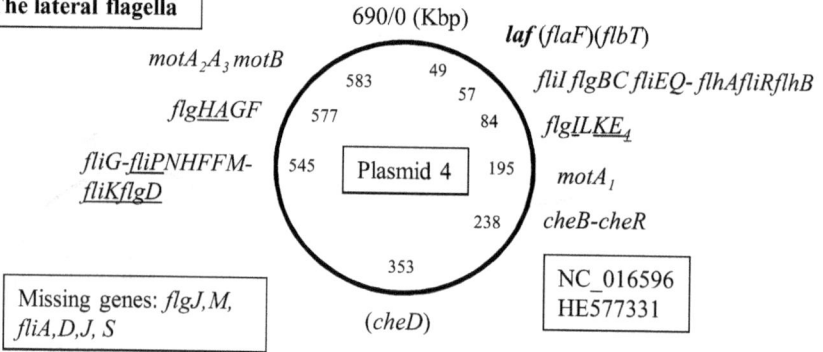

The lateral flagella

690/0 (Kbp)

laf (*flaF*)(*flbT*)

motA₂A₃ motB

flgHAGF

fliG-fliPNHFFM-fliKflgD

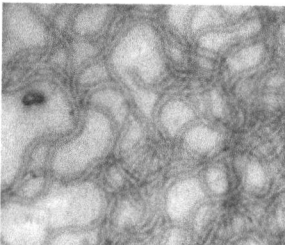

Plasmid 4

583   49
577   57
545   84
              195
              238
       353

flil flgBC fliEQ- flhAfliRflhB

flgILKE₄

motA₁

cheB-cheR

NC_016596
HE577331

Missing genes: *flgJ,M, fliA,D,J, S*

(*cheD*)

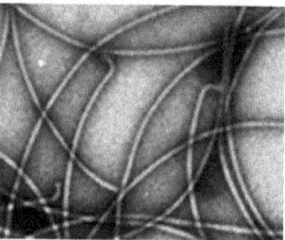

Purified **lateral flagella** give rise to a single band by SDS-PAGE, whose apparent molecular size is ~43 kDa. The lateral flagellin is not glycosylated. Judging from the molecular size and the location of the gene, Laf (417 aa) on plasmid 4 might be the lateral flagellin.

The basal body of the lateral flagella (lower panel). There are four Mot proteins for the lateral flagella: three MotA₁₂₃ and one MotB on plasmid 4. How these proteins are integrated in the motor is not clear.

Urgent proposal for interesting projects

---

1. Flagellar proteins should not have a signal peptide!

Most flagellar proteins are secreted without cleavage through the type III secretion system. Only FlgA, FlgH, and FlgI have signal peptides and are secreted through the Sec system into the periplasmic space. However, the genes underlined in the genetic map are predicted to retain signal peptides: *flgE$_1$K* on plasmid 1, *flgD*, *flgE$_2$*, *flgK*, and *fliD* on plasmid 3, and *flgD*, *flgKE$_4$*, *fliK*, and *fliP* on plasmid 4. For example, six cleavage sites (*//*) are predicted for FlgE$_4$ (410 aa) on plasmid 4.

N-terminus: MSLFSAMRSGVSGMSAQS//SR//M//A//AI//S// DNISNSATIG---C-terminus

If this is the case, this will cause a serious problem in flagellum assembly and motor function (see Chapter 32, *M. voltae*).

2. What is the role of the bushy hook?

The hook of the polar flagellum appears to be covered with bushy materials. There are three *flgE* candidate genes. FlgE$_1$ on plasmid 1 is extraordinarily large (1144 aa), while the others are ordinary sizes (E$_2$: 430 aa, E$_3$: 456 aa). Judging from the size and position, FlgE$_1$ would be the polar hook protein. Whether the bushy hook is derived from the large molecular size of FlgE$_1$ or from glycosylation of the other FlgE awaits direct proof.

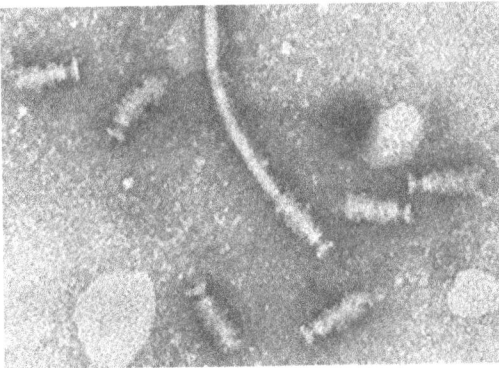

Bushy hooks remain intact (as they were) even after acid treatment, which is employed to remove filaments. The role of the bushy hook is not clear.

Strains were provided by Gladys Alexandre of Georgia Institute of Technology, Atlanta, USA.

The project was carried out by Masaomi Kanbe, Tatsuro Ebisawa, and Bonnie Jennifer.

CHAPTER 4

Bacillus subtilis — The Representative of Gram-Positive Bacteria

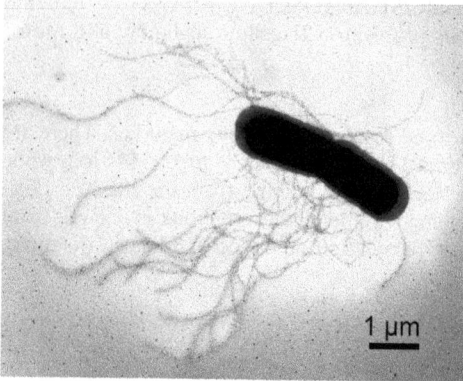

Phylum	Firmicutes
Class	Bacilli
Order	Bacillales
Family	Bacillaceae
Genus	Bacillus
Species	*B. subtilis*

Bacillus subtilis is one of **the best-characterized bacterium** among Gram-positives and is regarded as a model system for cell differentiation and development; *B. subtilis* cells produce spores under adverse conditions. In vegetative form, *B. subtilis* flagella are growing at apparently peritrichous positions, but in a more ordered way (see next page).

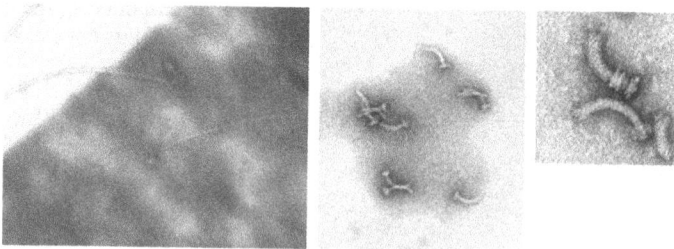

One of the distinctive differences in the flagellar structure of Gram-positive and Gram-negative species is **the absence of the PL ring complex** in the former.[1] There appear to be no ring structures corresponding to the PL ring in intact flagella on the cell wall (left) of an osmotically-shocked cell (see Appendix) or in the isolated HBB (middle) in *B. subtilis*.[2] Compare the structures of the hook-basal body between *B. subtilis* (arrow) and *Salmonella* in the same preparation (right). The fact that *B. subtilis* flagella rotate as smoothly as *Salmonella* flagella casts a doubt on the assumption that the PL ring works as the bushing for sustaining torque from the motor. It is more likely that the PL ring is necessary just for the hook to penetrate the outer membrane of Gram-negatives (see Introduction).

Bacillus subtilis subsp. subtilis str. 168 genome (4,215,606 bp/ 4,176 genes)	NC_000964 AL009126

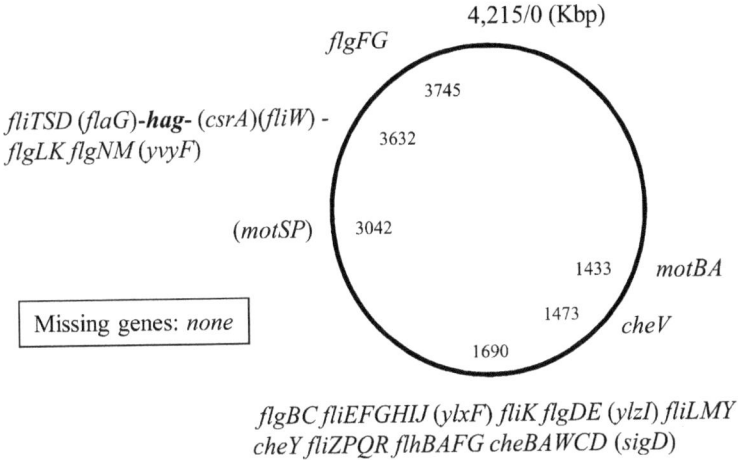

4,215/0 (Kbp)

flgFG

3745

3632

*fliTSD (flaG)-**hag**- (csrA)(fliW) -*
flgLK flgNM (yvyF)

(motSP)   3042

1433   *motBA*

1473   *cheV*

1690

Missing genes: *none*

flgBC fliEFGHIJ (ylxF) fliK flgDE (ylzI) fliLMY
cheY fliZPQR flhBAFG cheBAWCD (sigD)

The genome of *Bacillus subtilis* subsp. subtilis str. 168 strain consists of a single 4.2-Mbp chromosome. The flagellar genes essential for assembling the basal structure form a compact cluster at 1,690 bp on the chromosome, while the late genes for HAPs and the filament are found at 3,632 bp.[3] There are tentative gene names specific to *B. subtilis*: *ylx*, *ylz*, etc., which will be replaced by functional gene names in the future. The functional unit of the genes (*flgAHIJ*) required to form the PL ring complex is missing in accordance with EM observations (see Introduction: Functional units). Hag is the original name for flagellin or FliC and is used only for *Firmicutes*, *Actinobacteria*, and *Cyanobacter* species. The flagellar gene regulation of *B. subtilis* is a little different from that of *S. typhimurium*.[4] The *sigD* gene encodes the flagellum specific sigma factor (corresponding to *fliA*). A pair of *flhFG* genes that place a flagellum at a pole in some species is used to place the basal bodies in a grid-like pattern symmetrically around the cell body,[5] distinguishing them from the peritrichous position (see Topic: Flagellar position and shape).

There are four mot genes: *motA*, *motB*, *motS*, and *motP*.[6] MotA–MotB complex is the stator of the proton-driven motor, while MotP (a MotA homolog), together with MotS (a MotB homolog) forms the stator of the sodium-driven motor (see Topic: Mot proteins). Both stators may be used under different ion conditions, though this assumption awaits direct proof. The C rings (left), the rotors of the motor, are evident in this crude preparation.

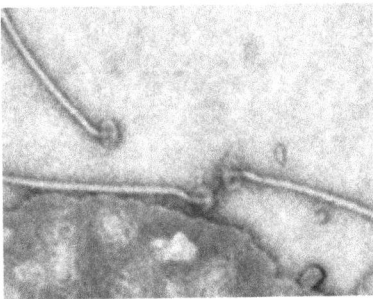

Strains were provided by George Ordal of the University of Illinois at Urbana–Champaign, USA.

Gene Regulation

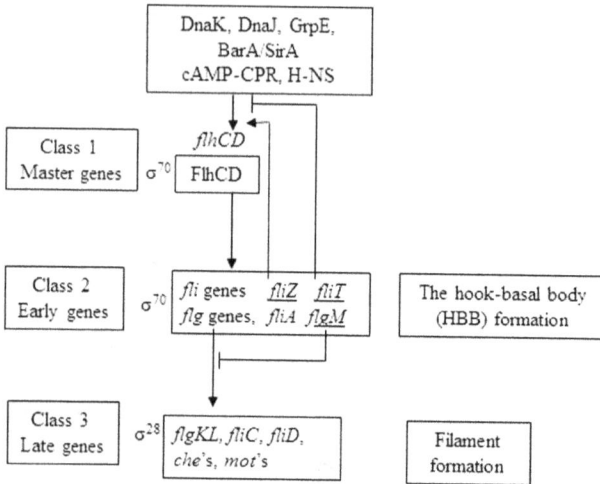

Gene regulation for peritrichous flagella.

Master operon: In *S. typhimurium* and related species, there is a regulatory hierarchy that governs the transcription of the flagellar genes of 3 classes.[1] The master operon (*flhDC*) is a sole operon in class 1 and is transcribed by the "housekeeping" sigma factor 70 (σ^{70}).[2,3] The sequences upstream of the *flhDC* operon are fairly diverse among bacterial species. In *E. coli*, the *flhDC* expression is affected by several genes and physiological factors: the heat shock proteins (DnaK, DnaJ, and GrpE),[4] the pleiotropic response regulator (OmpR) activated by acetyl phosphate,[5] and the DNA-binding protein H-NS. More such genes and factors have been discovered in other species.

Positive and negative regulators: With the help of σ^{70}, the master operon *flhDC* activates the class 2 genes that mostly code for structural proteins of the hook-basal body (HBB).[6] The class 2 operons contain four regulatory genes: *fliA*, *fliZ*, *fliT*, and *flgM*. FliZ and FliT are positive and negative, respectively, regulators for the class 2 operons, resulting in a balanced expression of the HBB components.[7] FliA is a sigma factor 28 (σ^{28}) for initiating transcription of the class 3 operons, and FlgM is an anti-sigma factor that tightly binds σ^{28} to halt its action.[8,9] FlgM is secreted through the central channel of the HBB completed by the action of FliK.[10,11] resulting in release of σ^{28}, which can then freely interact with RNA polymerase and direct transcription of the class 3 operons. All of these regulatory genes are transcribed from the class 2 as well as the class 3 operons. The amount of FlgM expressed at the class 3 is much higher than that at class 2, indicating an autogenous regulation of the class 3 operons.[6] Upon initiation of the class 3 gene expression, flagellar filaments assemble and the sensory system is organized. At the same time, the MotA/B complexes assemble around the motor, thereby completing a functional flagellum.[12]

The gene regulation just mentioned for *S. typhimurium* is a paradigm of the flagellar gene regulation, which is representative of all species with minor modifications. However, an important factor is missing in the scheme, which is FliK. Without FliK, the secretion gate does not allow FlgM to go out and subsequently suppresses the expression of the late genes. I added FliK in new schemes as one of the regulators. The molecular mechanism—the way in which FliK switches the substrate specificity—is still unknown (see Topic: Hook length).

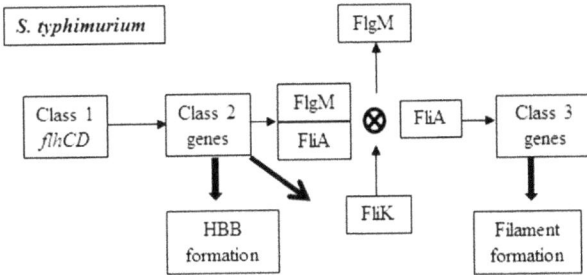

Gene regulation for polar flagella.

Polarly-flagellated species have a regulation mechanism for flagellar gene expression different from that of *S. typhimurium*. In *P. aeruginosa*, flagellar genes are divided into 4 classes in the transcriptional hierarchy.[13] Class 1 contains a sole *fleQ* gene that encodes the σ^{54}-dependent activator for the class 2 operons.[14] Most of the class 2 genes are used to construct the cytoplasmic components of the HBB, which is a part of the export apparatus. Class 2 also includes three regulatory genes: *fleS*, *fleR*, and *flgM* (anti-σ^{28}).[14] FleS activates expression of the class 3 genes, which encode the hook protein and HAPs. FliA activates the class 4 genes, which encode filament components and motor proteins. Here again, FliK is an important regulation factor to open the secretion gate for FlgM.

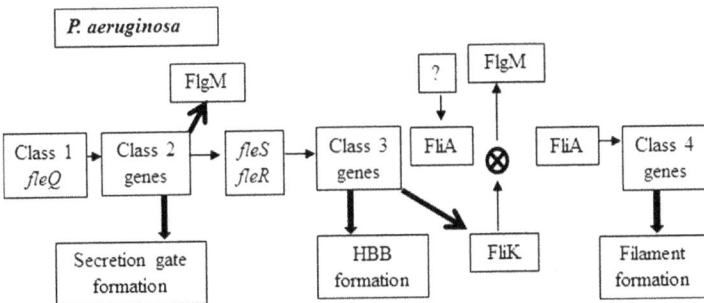

CHAPTER 5

Bdellovibrio bacteriovorus — A Small but Fierce Predator

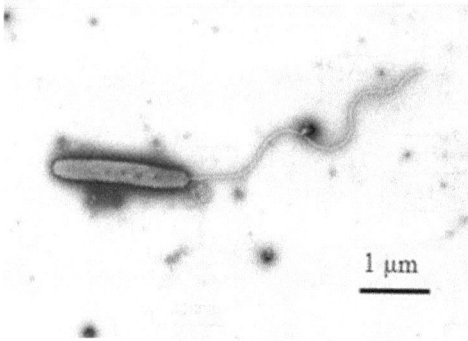

Phylum	Proteobacteria
Class	Delta Proteobacteria
Order	Bdellovibrionales
Family	Bdellovibrionaceae
Genus	Bdellovibrio
Species	B. bacteriovorus

Bdellovibrio bacteriovorus is a Gram-negative, aerobic/microaerophilic bacterium that **preys upon a wide variety of other Gram-negative** bacteria. Its life cycle has two major stages: a free-swimming stage spent searching for prey in water or soil (the "attack phase") and a growth stage spent **inside the prey's periplasmic space** between its inner and outer membranes.[1] After colliding with a prey cell and irreversibly attaching to its surface, the predator breaches the outer membrane, stays in the periplasm, and grows by an **uneven cell division**. Each cell has a single, polar, sheathed flagellum and swims fast in water.[2]

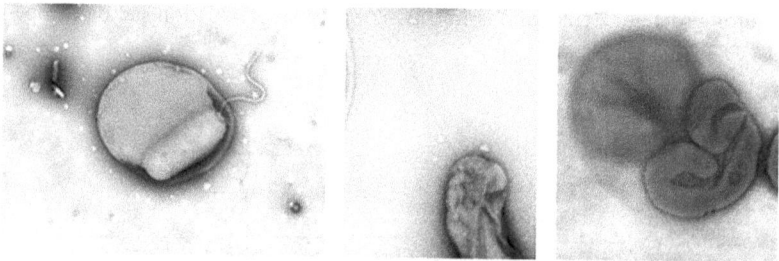

A *B. bacteriovorus* cell invades the periplasmic space of the host *E. coli* cell (left), using a type IV pili[3] (middle). Cells in the host periplasm (bdelloplast), multiply not in a binary manner, but in an uneven manner (right). In the bdelloplast, flagella disappear.

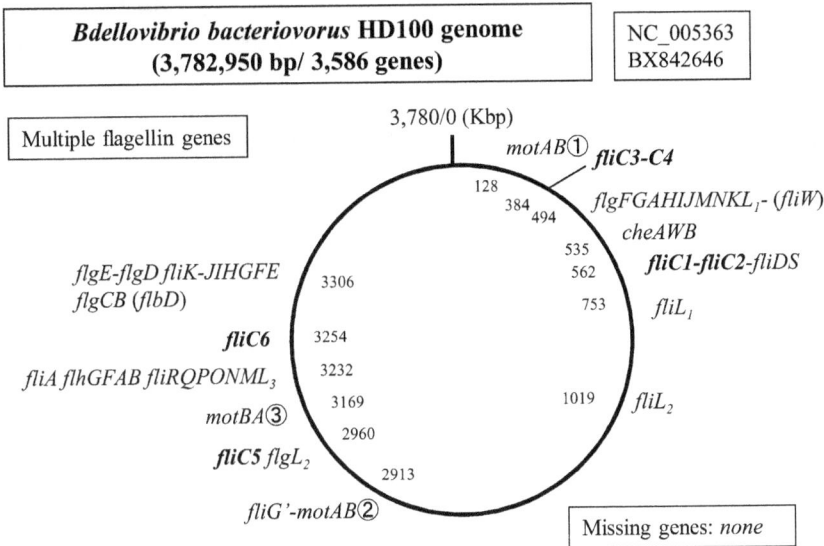

Bdellovibrio bacteriovorus HD100 genome **(3,782,950 bp/ 3,586 genes)**	NC_005363 BX842646

Multiple flagellin genes

3,780/0 (Kbp)

*motAB*① *fliC3-C4*
128
384 494   *flgFGAHIJMNKL₁- (fliW)*
535   *cheAWB*
562   *fliC1-fliC2-fliDS*
753   *fliL₁*

flgE-flgD fliK-JIHGFE
flgCB (flbD)
3306
3254
fliC6
fliA flhGFAB fliRQPONML₃
3232
3169   1019   *fliL₂*
*motBA*③
2960
fliC5 flgL₂
2913

*fliG'-motAB*②

Missing genes: *none*

The genome of *B. bacteriovorus* HD100 strain consists of a single 3.7-Mbp chromosome. Flagellar genes are scattered at several loci all over the chromosome. There are six flagellin genes (see Topic: Multiple flagellins). There are three sets of *mot* genes[4]: all of them are the stators of a **proton-driven motor**, although one of them (*motAB2*) shows a sequence homology with that of a **sodium-driven motor** (see Topic: Mot proteins). The function of three *fliL* genes is not known.

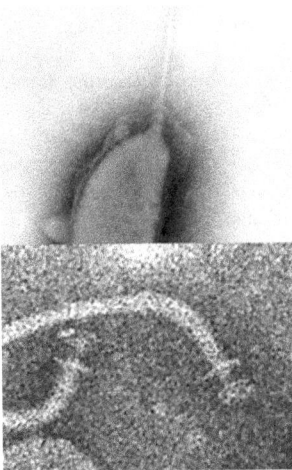

The LP ring complex that is connected to the inner and outer membranes is not distinctive as much as those in other Gram-negative bacteria. The outermost L ring is seen, but the P ring underneath the L ring is not visible. Because there exists a *flgI* gene for the P ring protein, the P ring structure could be smaller than usual, or it could have been destroyed during purification.

Strains were provided by Elizabeth Sockett of Nottingham University, Nottingham, UK.
The project was carried out by Yoshiko Iida and Kaoru Uchida.

CHAPTER 6

Borrelia burgdorferi — Periplasmic Flagella in a Flat Wave Body

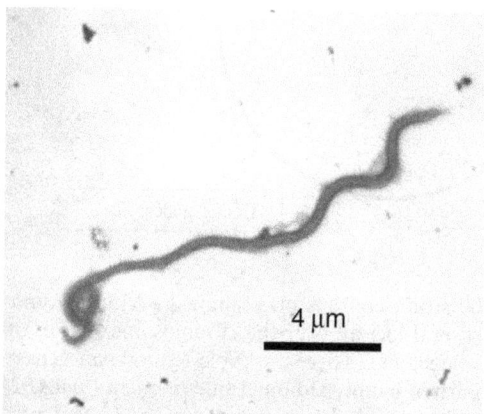

4 μm

Phylum	Spirochaetes
Class	Spirochetes
Order	Spirochaetales
Family	Spirochetaceae
Genus	Borrelia
Species	*B. burgdorferi*

Borrelia burgdorferi is a spirochete found in deer/bear ticks as the causal agent of Lyme disease. **The cell is flat-wave shaped** and Gram-negative. Each cell possesses **endoflagella or periplasmic flagella** which are confined within the periplasmic space. One of the most striking features of *B. burgdorferi* as compared with other eubacteria is its unusual genome, which consists of **a linear 0.9-Mbp chromosome and numerous linear and circular plasmids**.

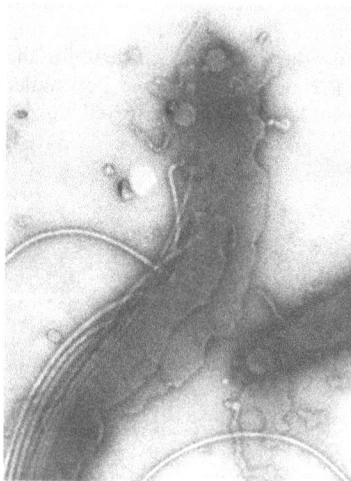

Flagella grow from near the cell poles and are aligned in an orderly manner along the cell axis. The filament contains two kinds of flagellins: FlaA and FlaB. FlaA is the minor flagellin of 36 kDa and localizes at the proximal end of the filament. FlaB is the major flagellin of 42 kDa and occupies the rest of the filament. Flagellin synthesis is not controlled by sigma 28, but by sigma 70.

Wild-type filament isolated from cells is a mixture of two types of helices: one with a pitch of 1.88 μm (10%) and the other with a pitch of 1.42 μm (90%). In the absence of FlaA, filaments are all in a short-pitch helix.

Borrelia burgdorferi B31- genome
(1,521,208 bp/ 1,346 genes)

NC_001318

Multiple flagellin genes

flhGFAB fliRQPZYML motBA (flbD) flgED fliK(flbB)(flbA) fliIHGFEflgCB
←——

(*flaA') flgKL (fliW)* | *flaA cheAWXY* | *flhG*
182 | 281 | 705 | 764 | | 910
0 | | | | | (Kbp)
147 | 226 | 561 | 812

flaB fliD | *fliG'* | *fliS cheY* | *(rpoS)-flgI-flgGF*

Missing genes: *flgM, fliAJ*

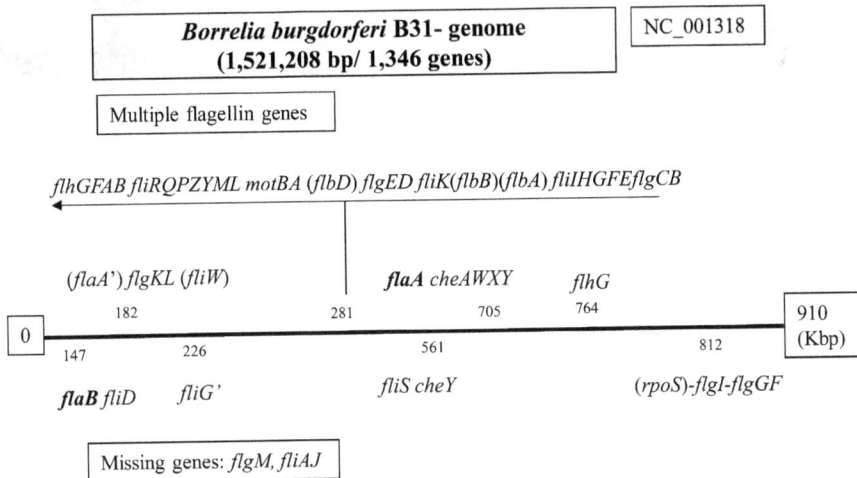

The genome of *Borrelia burgdorferi* B31 strain consists of one 910,724 bp **linear chromosome, 12 linear plasmids, and 9 circular plasmids**. The majority of the flagellar genes are localized in a locus at 281 Kbp on the chromosome, and the rest are scattered at several loci on the linear chromosome. FlgI, the component of the P ring, is missing but is not necessary for motility in this organism.

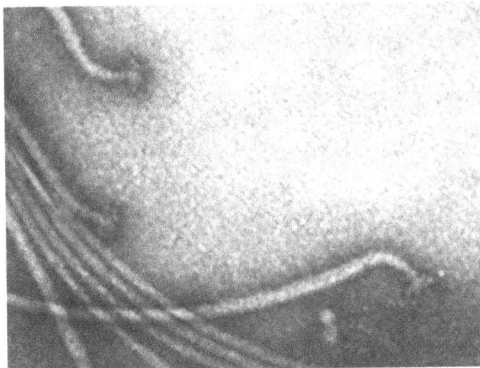

Although *B. burgdorferi* cells are **Gram-negative**, the flagellar basal body **does not retain the PL ring**, as is present in the Gram-positive bacteria. This is because the flagella grow in the periplasmic space and do not penetrate the outer membrane.

The hook protein FlgE has a predicted MW of 48 kDa, but runs in a high MW region (>160 kDa) by SDS-PAGE, suggesting that the hook might be covalently cross-linked after polymerization. This is impossible for most species, but it can possibly happen in the periplasmic space of this organism.

Strains were provided by Nyles Charon of West Virginia University, Morgantown, USA.
The project was carried out by Satoshi Shibata, M. Abdul Motaleb, and Chunhao Li.

CHAPTER 7

Bradyrhizobium japonicum — The Nitrogen–Fixer in the Peanut Farm

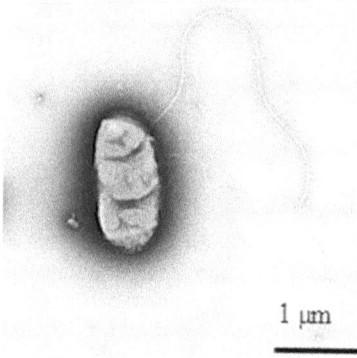

1 μm

Phylum	Proteobacteria
Class	Alphaproteobacteria
Order	Rhizobiales
Family	Bradyrhizobiaceae
Genus	Bradyrhizobium
Species	*B. japonicum*

Bradyrhizobium japonicum is a nitrogen-fixing bacterial species that forms root nodules specifically on soybean (*Glycine max*) roots. Soybean has long been the most popular and important protein source in Japan. Each cell possesses two types of flagella: one **thick flagellum** and a few **thin flagella**, both types of which originate from the same area of **the subpolar region**. Both flagella are powered by the proton motive force. *B. japonicum* also has type III secretion systems (T3SS), which play a crucial role in the plant–microorganism interactions, especially for bacterial adhesion to root hair surfaces and symbiotic incompatibility.

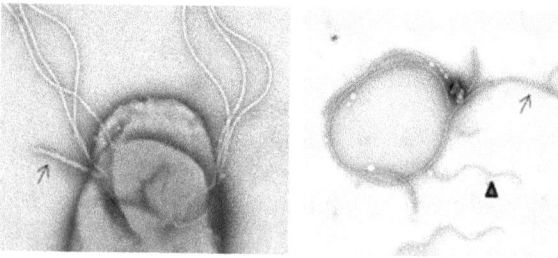

Two types of flagella in *B. japonicum* cells are unusual. They look like polar flagellum and lateral flagella as seen in *Vibrio* species, but we should not call them by the conventional names, because of their growing positions at the subpolar region (left panel). Only the thick flagellum (arrow) grows in the tryptone/yeast extract/cystine (TYC) medium, while both flagella grow in the minimal medium (MM) (see Chapter 3, *A. brasilense*). The thick flagellum (arrow) has a larger pitch than the thin flagellum (arrow head), similar to those of the polar and the lateral flagellum (right panel).

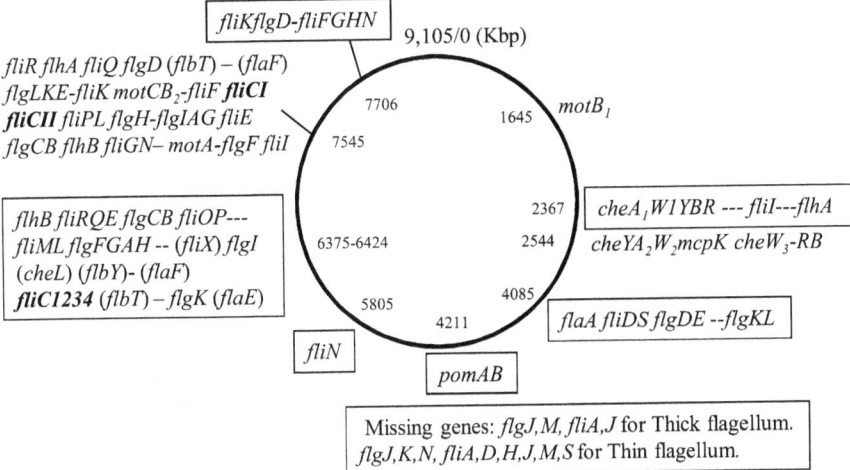

Bradyrhizobium japonicum USD110 genome
(9,105,828 bp/ 8,317 genes)

NC_004463
BA000040

Multiple flagellin genes

fliKflgD-flifFGHN

9,105/0 (Kbp)

fliR flhA fliQ flgD (flbT) – (flaF)
flgLKE-fliK motCB₂-fliF fliCI
fliCII fliPL flgH-flgIAG fliE
flgCB flhB fliGN– motA-flgF fliI

7706

7545

1645   *motB₁*

flhB fliRQE flgCB fliOP---
fliML flgFGAH -- (fliX) flgI
(cheL) (flbY)- (flaF)
fliC1234 (flbT) – flgK (flaE)

6375-6424

2367
2544

cheA₁W1YBR --- fliI---flhA
cheYA₂W₂mcpK cheW₃-RB

5805

4085

4211

fliN

pomAB

flaA fliDS flgDE --flgKL

Missing genes: *flgJ,M, fliA,J* for Thick flagellum.
flgJ,K,N, fliA,D,H,J,M,S for Thin flagellum.

The genome of *Bradyrhizobium japonicum* USD110 strain consists of a single 9-Mbp chromosome. The genome is one of **the largest among eubacteria**. Flagellar genes are clustered at around 6,400 Kbp for the thick flagellum (in box) and at around 7,545 Kbp for the thin flagellum. There are six flagellins: four flagellins (FliC1234: 756–763 aa) are for the thick filament, and two flagellins (FliCI: 314 aa and FliCII: 313 aa) are for the thin filament. FliCI and FliCII are 90% homologous. Flagellar gene arrangement in both systems is random in a manner similar to *Caulobacter crescentus*. The mechanism for placing both types of flagella at the subpolar region is not known; a pair of *flhFG* genes, which are necessary for placement of the polar flagellum, has not been found.

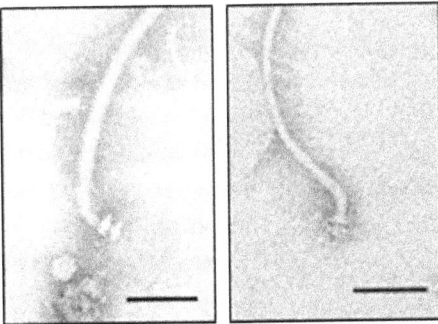

The thick flagellum is more stable against acid, but more labile to heat than the hook. Thus, for a preparation of HBB, heat treatment is employed. The molecular size of flagellins is 65 kDa for the thick filament (left) and 33 kDa for the thin filament (right). Bars indicate 100 nm.

Strains were provided by Kiwamu Minamizawa of Tohoku University, Sendai, Japan, and by Michael Goettfert of Dresden University of Technology, Dresden, Germany.

CHAPTER 8

Caulobacter crescentus — Alteration between Flagellum and Stalk

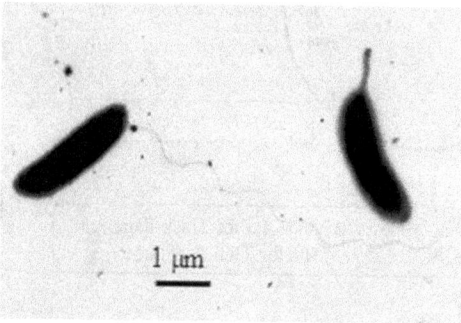

Phylum	Proteobacteria
Class	Alphaproteobacteria
Order	Caulobacterales
Family	Caulobacteraceae
Genus	Caulobacter
Species	*C. crescentus*

Caulobacter crescentus (or *C. vibrioides*) is a Gram-negative bacterium that grows in aquatic environments. *C. crescentus* cells invariably differentiate and divide asymmetrically at each cell cycle to produce two types of cells: **a swimming cell with a polar flagellum** (left) and **a sessile cell with a stalk** (right) that sticks to a solid surface. It is a simple single-celled model system in which to study cellular differentiation.

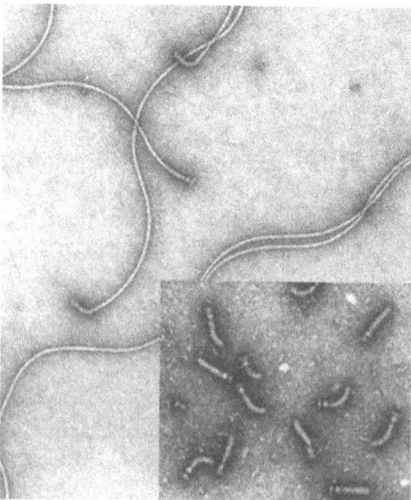

C. crescentus flagella (left) appear similar to those of *Salmonella*. The hook (591 aa) is thicker than the *Salmonella* hook (402 aa).
Ring structures (below) are often observed in the membrane fraction.

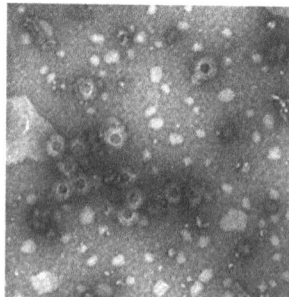

```
┌─────────────────────────────────────────────┐   ┌─────────────┐
│  Caulobacter crescentus NA1000 genome         │   │ NC_011916   │
│       (4,042,929 bp/ 3,877 genes)             │   │ CP001340    │
└─────────────────────────────────────────────┘   └─────────────┘
```

Multiple flagellin genes

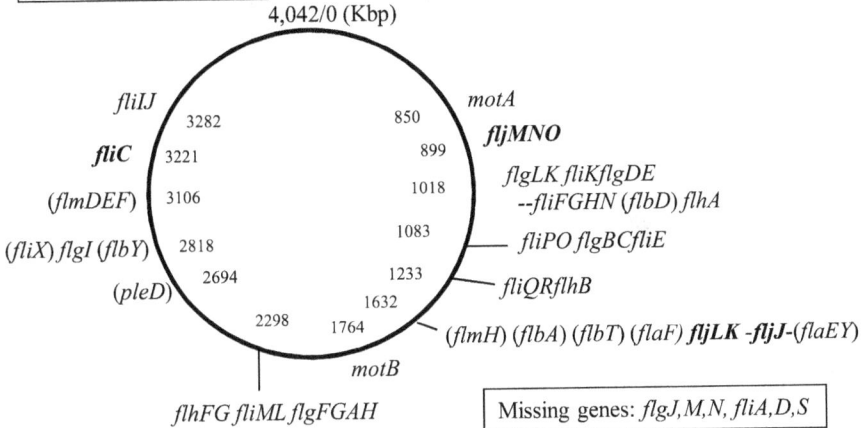

4,042/0 (Kbp)

fliIJ   850   *motA*
          3282   *fljMNO*
fliC   3221   899
                            1018   *flgLK fliKflgDE*
(*flmDEF*) 3106   --*fliFGHN* (*flbD*) *flhA*
                            1083
(*fliX*) *flgI* (*flbY*) 2818   —— *fliPO flgBCfliE*
          2694   1233
(*pleD*)   1632   — *fliQRflhB*
      2298   1764
                      (*flmH*) (*flbA*) (*flbT*) (*flaF*) *fljLK* -*fljJ*-(*flaEY*)
              motB

flhFG fliML flgFGAH   ┌──────────────────────────────────────┐
                               │ Missing genes: *flgJ,M,N, fliA,D,S*   │
                               └──────────────────────────────────────┘

The genome of *Caulobacter crescentus* NA1000 consists of a single 4-Mbp chromosome. Flagellar genes are scattered all over the chromosome, forming small clusters of genes, similar to those of *H. pylori* or *B. bacteriovorus*. The gene arrangement is similar to that of *M. magnetotacticum*. There are **seven flagellin genes** (*fljJ–fljO, fliC*) in this strain. Six flagellins (FljJ–O) are homologous and have a similar size (273–276 aa), while FliC is larger than the others, 424 aa. The six flagellins are incorporated into a filament (see Topic: Multiple flagellins). Whether FliC is included in the filament is not known. The *flm* genes are involved in protein glycosylation.

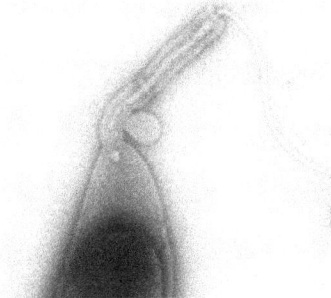

The polar flagellum and the stalk are not exclusive in a *pleD* mutant. PleD is a diguanylate cyclase and is involved in flagellum ejection. In this mutant, the flagellar motor is confined to the smallest area ever observed.

Strains were provided by Urs Jenal of Basel University, Basel, Switzerland.

CHAPTER 9

Enterococcus casseliflavus — **Edible Flagella**

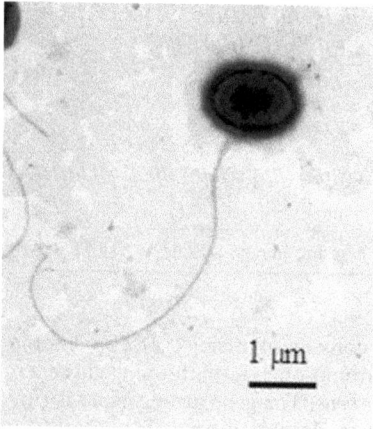

1 μm

Phylum	Firmicutes
Class	Bacilli
Order	Lactbacillales
Family	Enterococcaceae
Genus	Enterococcus
Species	*E. casseliflavus*

Enterococcus casseliflavus is a Gram-positive, facultative anaerobic, lactic acid bacterium. *E. casseliflavus* is one of the probiotics that are used to make yogurt as a health food. Although *E. faecalis* in the same genus is often referred to as an opportunistic pathogen, it is also commonly used in probiotic preparation. It should be noted that even popular probiotic lactic-acid bacteria such as *Bifidobacterium* or *Lactobacillus* may also cause septicemia (a blood poisoning) to immune-depressed persons.

Enterococci species are unable to form spores, but instead cells are very resistant to drying and tolerant of a wide range of environmental conditions: extreme temperature (10–45°C), pH (4.5–10.0) and high sodium chloride concentrations.

E. casseliflavus cells are flagellated, which is rare among the genus *Enterococci*. Each cell possesses a single polar flagellum. The flagellin may be recognized by TLR5 to trigger innate immune responses (see Topic: Flagellin size).

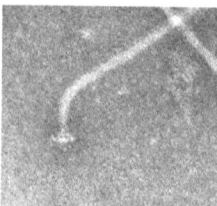

The cell wall of *E. casseliflavus* is very hard to digest by lysozyme in accordance with its tolerance against harsh conditions, as mentioned above. The basal body isolated from occasionally-lysed cells shows only two rings—as expected—for the Gram-positive bacteria.

<table>
<tr><td>

***Enterococcus casseliflavus* EC20 genome**
(3,427,276 bp/ 3,114 genes)

</td><td>

NC_020995
CP004856

</td></tr>
</table>

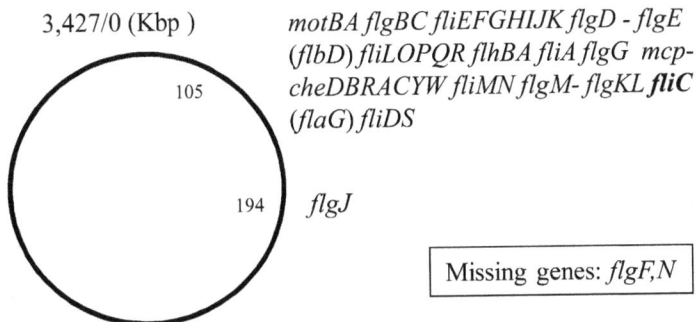

3,427/0 (Kbp)

motBA flgBC fliEFGHIJK flgD - flgE
(flbD) fliLOPQR flhBA fliA flgG mcp-
*cheDBRACYW fliMN flgM- flgKL **fliC***
(flaG) fliDS

flgJ

Missing genes: *flgF,N*

The genome of *Enterococcus casseliflavus* EC20 strain consists of a single 3.4-Mbp chromosome. Flagellar genes form a compact cluster at 105 kbp. This is one of **the most compactly packed clusters**. The gene arrangement is similar to that of *B. subtilis*. Naturally, *E. casseliflavus* FliC (360 aa) is 47% homologous with *B. subtilis* FliC (305 aa).

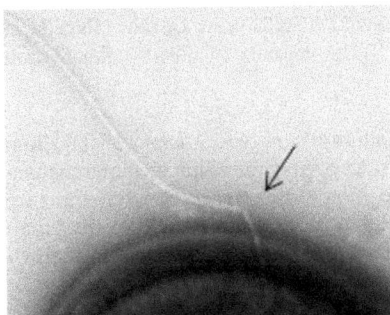

Purified filament samples gave a single band of flagellin at 43 kDa in SDS-PAGE, which is a little larger than the predicted molecular size (360 aa). The hook of *E. casseliflavus* often shows a wavy shape and is evident on the cell (arrow). This may suggest polymorphic transition at neutral pHs, as seen in *S. ruminantium* (see Chapter 25).

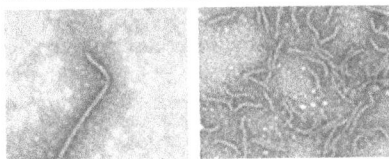

Strains were provided by Takashi Shimada of Nichinichi Pharmaceutical Co., Ltd. The project was carried out by Kaoru Uchida.

CHAPTER 10

Escherichia coli — The Representative of the Gram-Negative Bacteria

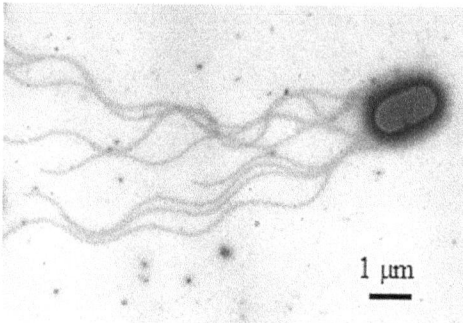

1 μm

Phylum	Proteobacteria
Class	Gammaproteobacteria
Order	Enterobacteriales
Family	Enterobacteriaceae
Genus	Escherichia
Species	*E. coli*

Escherichia coli is one of the premier model organisms used in studies of bacterial genetics, physiology, biochemistry, and so on. **Studies on motility and chemotaxis have been most actively done** with *E. coli* strains,[1–5] whereas there have not been as many studies on flagellar structures with *Salmonella* strains. *E. coli* cells produce more flagella in a low-salt medium than in a high-salt medium.[6] In the presence of glucose, flagellation is suppressed due to **catabolite repression**.

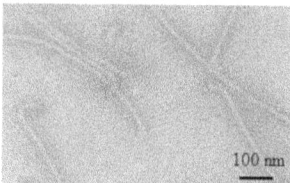

100 nm

The *E. coli* flagellum ressembles *Salmonella* flagellum in many aspects, and their flagellins are interchangeable.[7] There are minor differences between the two. *E. coli* filament is more sensitive to acid or heat than *Salmonella* filament. *Salmonella* filament is methylated, while *E. coli* filament is not.

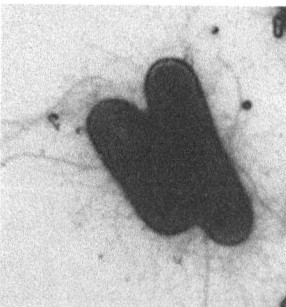

E. coli cells grow flagella and pili alternatively depending upon environmental conditions. On a cell, pili overgrow flagella and vice versa. Both structures are involved in developing pathogenicity.

```
┌──────────────────────────────────────────┐   ┌──────────────┐
│  Escherichia coli K-12 W3110 genome        │   │  NC_007779   │
│  (4,646,332 bp/ 4,217 genes)               │   │              │
└──────────────────────────────────────────┘   └──────────────┘
```

4,646/0 (Kbp)

$\overrightarrow{tsr}$ 4596

1130

| Region I |

flgNMA *flgBCDEFGHIJKL*

3216
aer

1494 $\overrightarrow{trg}$

2014   1960 | Region II |

2001

flhEAB cheZYBR tap tar cheWAmotBA flhCD

fliE FGHIJK LMNOPQR

| Region IIIB |

fliYZA fliC D ST | Region IIIA |

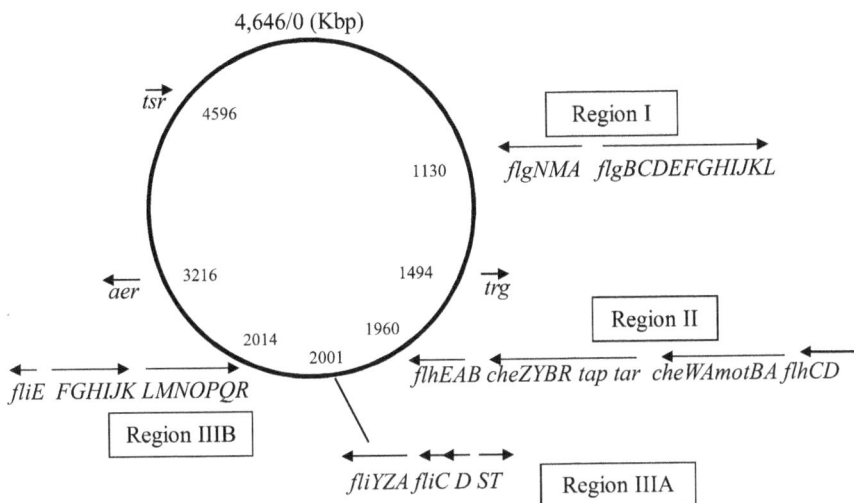

The genome of *Escherichia coli* K-12 W3110 strain consists of a single 4.6-Mbp chromosome. Flagellar genes are clustered in four regions in the same manner as in *Salmonella* (see Chapter 24). **Five major MCPs** are indicated by their special names: **trg**, **tap**, **tar**, **aer**, and **tsr** (see refs. 25–27 in Introduction).

The *E. coli* genome is more flexible than the *Salmonella* genome in the gene transfer and can vary by more than 1 Mbp of DNA through genomic expansion, deletion, and rearrangement to produce the **pathogenic E. coli clans.**[8,9]

Genome Sizes in Pathogenic *E. coli*			
	Chromosome	Plasmids	Total Nclts
E.coli K12	4646332	0	4646332
Enteropathogenic E.coli (EPEC)/O127:H6	4965553	2	5069678
Extraintestinalpathogenic E. coli (ExPEC)	5132068	0	5132068
Enteroadherent E.coli (EAEC)	5154862	0	5154862
Uropathogenic E.coli (UPEC)/O6:K2:H1	5231428	0	5231428
Enterotoxigenic E.coli (ETEC)/O139:H28	4979619	6	5249288
Enterotoxigenic E.coli (EHEC)/O78:H11:K80	5153435	4	5325888
Enterotoxigenic E.coli (EHEC)/O157:H7	5528445	1	5620522
Enterotoxigenic E.coli (EHEC)/O26:H11	5697240	4	5855531

The pathogenic E. coli

There are several types of pathogenic *E. coli*, which include toxin-producing *E. coli*, attaching and effacing (A/E) *E. coli*, and invasive *E. coli*.

1. Toxin-producing *E. coli*

Enterotoxigenic *E. coli* (ETEC) O6:K15:H16 strain produces both heat-labile (LT) and heat-stable (ST) enterotoxins.

Enteroadherent *E. coli* (EAEC) produces the enteroaggregative ST (EAST).

Enterohemorrhagic *E. coli* (EHEC) O157:H7 strain produces Shiga-like toxin (Vero toxin).

2. Attaching and effacing (A/E) *E. coli*

Enteropathogenic *E. coli* (EPEC) O127:H6 and EHEC O157:H7 strains cause A/E lesions.

3. Invasive *E. coli*.

Enteroinvasive *E. coli* (EIEC) O112, O114 strains.

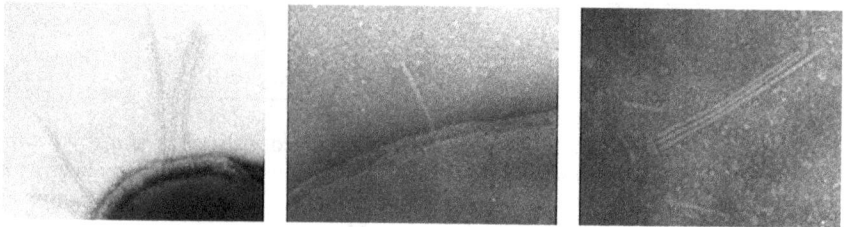

EspA filaments (left), the base on the EPEC cell (middle), and isolated EspA filaments with the base attached (right).

Enteropathogenic E. coli (EPEC) is a human pathogen. In addition to the flagella, the cells possess a type III virulence secretion apparatus, which appears similar to the needle of *Salmonella*, but is attached with a filament called EspA (see Topic: Flagella and pathogenicity). Thickness of the EspA filament is 12 nm, which measure is between the length of a flagellar filament (15 nm) and that of a type I pilus (9 nm).[10] How the EspA filament grows on the distal tip of the needle is not known.

Strains were provided by Gad Frankel of Imperial College London, UK. The project was carried out by Noriko Takahashi.

The second flagella system

Escherichia coli O44:H18 (EAEC) 042 genome (5,355,323 bp/ 4,920 genes)

NC_017626

5,241/0 (Kbp)

T3SS Virulence genes

284

flhAB fliRQPNM (lafK) fliEFGHIJ

3257

300

fliB flgNMA flgBCDEFGHIJKL (lafW) - - (lafZ) fliC fliDSTKLA motAB

chromosome

Flag-2

1219

2135-2184

Flag-1

Missing genes: *flhCD, fliO*

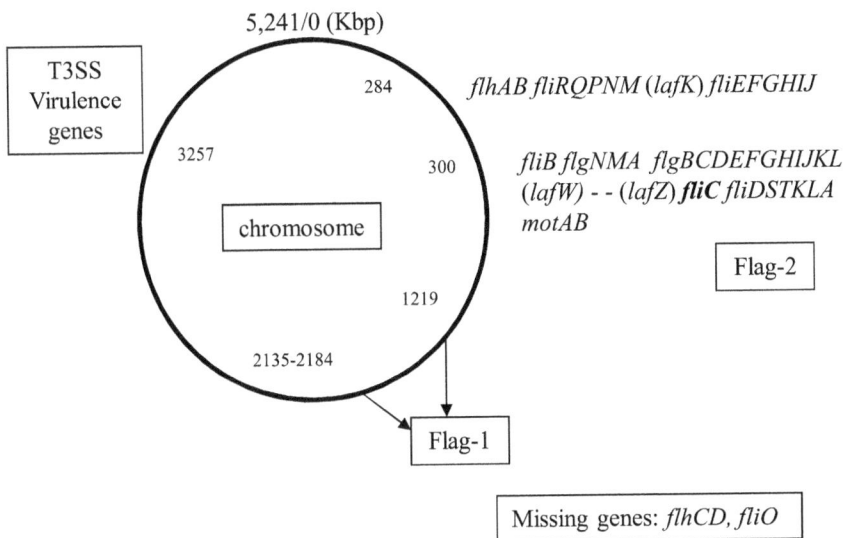

The genome of Enteroaggregative *E. coli* (EAEC) 042 strain consists of one 5.2-Mbp chromosome and one 113,346-bp plasmid.[11] The EAEC cell possesses two flagellar systems: Flag-1 and Flag-2.[12] The Flag-1 system produces actual flagella, while the Flag-2 system does not. The Flag-2 flagellum is cryptic in a similar manner to the *Shigella* flagellum (see Chapter 34). The Flag-2 genes are called new names (*lfg, lfi, lfh, etc.*) to distinguish them from the Flag-1 genes. Here, I simply use the old nomenclature for comparison with other flagellar systems.

The gene arrangements of the Flag-1 and Flag-2 systems are almost the same, but there are minor differences between the two. The Flag-2 lacks *fliO* but retains *fliB* (methylase for flagellin) and new genes with unknown function, *lafK, lafW,* and *lafZ.* The Flag-2 flagellin (304 aa) is 33% homologous with the Flg-1 flagellin (497 aa), but lacks the central region of the molecule.

Virulence genes (*epr, epa, eiv*) that are necessary for assembling the Type III virulence secretion system[13] are located in a locus at 3257 kbp.

CHAPTER 11

Geobacillus kaustophilus — The Heat- and Acid-Stable Flagella

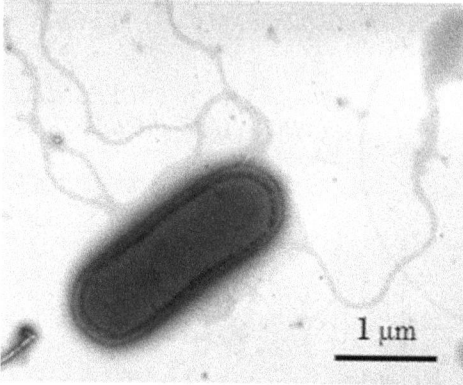

Phylum	Firmicutes
Class	Bacilli
Order	Bacillales
Family	Bacillaceae
Genus	Geobacillus
Species	*G. kaustophilus*

Geobacillus (formerly *Bacillus*) *kaustophilus* is an obligately **thermophilic bacterium** isolated from the deep-sea sediment of the Mariana Trench. The Gram-stain on this strain may vary between positive and negative. *G. kaustophilus* cells can grow in a wide range of temperatures, 35–75°C, with the optimal growth temperature in the lab being 60°C. Each cell possesses what seem to be peritrichous flagella, which might be regularly arranged around the cell body as seen in *B. subtilis* (see Chapter 4). The cell is surrounded by the S-layer.

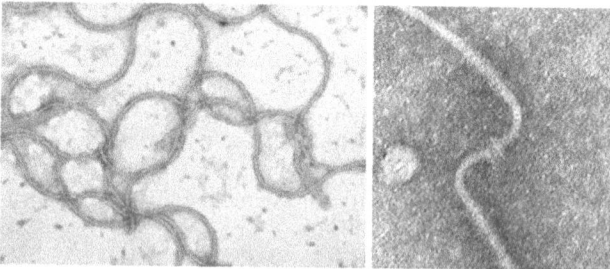

Filaments are stable against heat up to 60°C and against acid down to pH 3.5. The pitch of the filament is 1.27 μm. Purified filaments give rise to one band at 32 kDa by SDS-PAGE.

The flagellar basal body has the MS ring, rod, hook, HAPs, and filament; absence of the PL ring complex is common in Gram-positive bacteria.

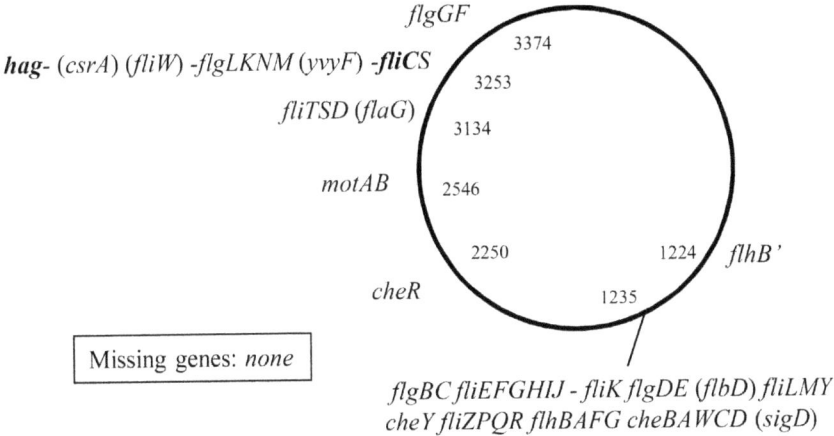

Geobacillus kaustophilus HTA426 genome (3,592,666 bp/ 3,539 genes)

NC_006510

Multiple flagellin genes

3,544/0 (Kbp)

flgGF

hag- (csrA) (fliW) -flgLKNM (yvyF) -fliCS

fliTSD (flaG)

motAB

cheR

3374
3253
3134
2546
2250
1224   *flhB'*
1235

Missing genes: *none*

flgBC fliEFGHIJ - fliK flgDE (flbD) fliLMY
cheY fliZPQR flhBAFG cheBAWCD (sigD)

The genome of the *G. kaustophilus* HTA426 strain consists of one 3.5-Mbp chromosome and one plasmid (47,890 bp). The gene arrangement is identical with those of *B. subtilis*. There are two flagellin genes: *hag* (297 aa) and *fliC* (604 aa). Comparing the molecular size, the filament should be composed of Hag flagellin.

Urgent proposal for an interesting project

The proximal portion of the filament appears thicker than the other part (arrow). This structure is new. A theory that the thick part might be composed of the large flagellin FliC should be tested. If this part is proved to be composed of the flagellin, this will be the first visible case of a localized flagellin in a filament (see Topic: Multiple flagellins).

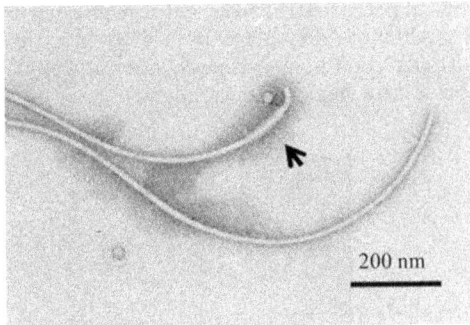

200 nm

Strain was provided by Toshiharu Yakushi of Yamaguchi University, Yamaguchi, Japan.
The project was carried out by Yuko Mozaki.

CHAPTER 12

Gluconobacter oxydans — The Vinegar Producing Bacteria

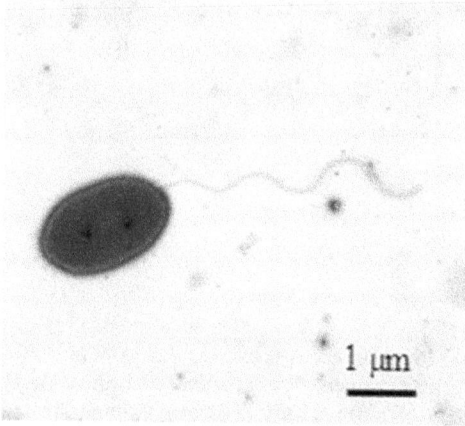

Phylum	Proteobacteria
Class	Alphaproteobacteria
Order	Rhodospirillales
Family	Acetobacteraceae
Genus	Gluconobacter
Species	*G. oxydans*

Gluconobacter oxidans (formerly *Acetobacter suboxydans*) is a Gram-negative oval-shaped bacterium. *G. oxydans* was found to be a species among bacteria with the ability to produce vinegar through alcoholic fermentation. *G. oxydans* causes apples and pears to rot. *G. oxydans* can oxidize ethanol to acetic acid but cannot utilize lactose as a carbon source, which makes this strain different from other acetic acid bacteria. Cells grow best at pH 5.0–6.0 at 30°C, but stop growing at temperatures above 37°C.

Although they are able to grow in extreme conditions, the growth rate in those extreme conditions is slow to reach to cell density high enough to observe. Each cell possesses a single polar flagellum.

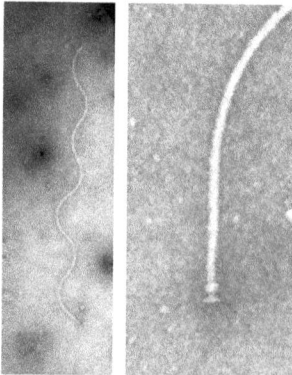

The flagellar basal body of *G. oxydans* resembles that of *Salmonella*. However, the hook is straight, as is often observed with polar flagella or with short hooks. Compare this hook with the short hooks of *X. oryzae* (Chapter 29).

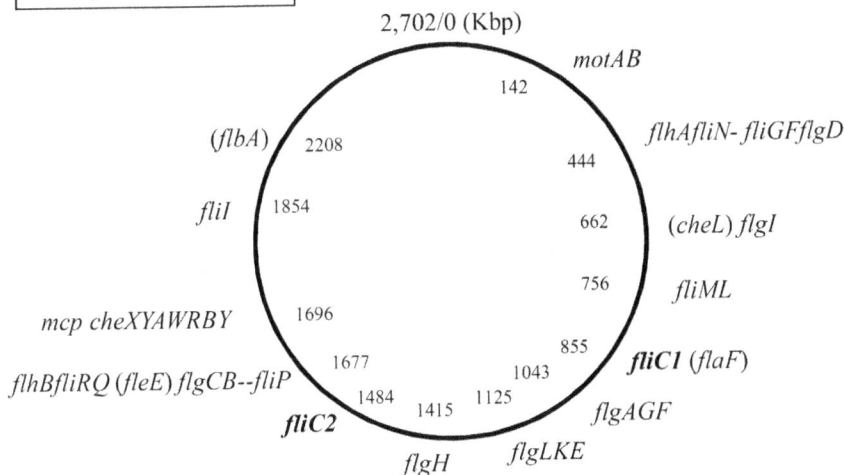

| **Gluconobacter oxydans 621H genome** **(2,922,384 bp/ 2,664 genes)** | NC_006677 |

Multiple flagellin genes

2,702/0 (Kbp)

motAB

142

(*flbA*)   2208

flhAfliN- fliGFflgD

444

fliI   1854

662   (*cheL*) *flgI*

756   *fliML*

mcp cheXYAWRBY   1696

855

flhBfliRQ (*fleE*) *flgCB--fliP*   1677   1043   *fliC1* (*flaF*)

1484   1415   1125   *flgAGF*

fliC2

flgH   *flgLKE*

Missing gens: *flgJ,M,N, flhFG, fliA,D,E,H,J,K,S*

The genome of *Gluconobacter oxydans* 621H strain consists of one 2,702,173bp chromosome, a megaplasmid (163 Kbp), and four plasmids (26.6, 14.5, 13.2, and 2.7 Kbp). *G. oxydans* has a small genome size due to its limited metabolic abilities.

Flagellar genes are scattered all over the chromosome as randomly as in *H. pylori*. Because of the randomness, it is not simple to annotate flagellar genes from their sequences and a relationship with neighbors, and there are so many genes to be found. There are two flagellins: FliC1 (491 aa) and FliC2 (792 aa), which is **one of the largest flagellins** (see Topic: Flagellin size). Compare with another large flagellin in *M. magnetotacticum* (Chapter 16).

A purified flagella sample gives rise to a major band at 48 kDa and a minor band at 85 kDa by SDS-PAGE.

The filament is stable down to pH 2.2 and does not show polymorphs at any pH values.

Strains were provided by Toshiharu Yakushi of Yamaguchi University.
The project was carried out by Manabu Konishi and Minori Matsubara.

CHAPTER 13

Helicobacter pylori — Randomly Arranged Flagellar Genes

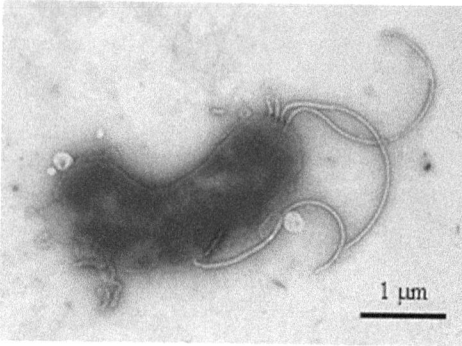

Phylum	Proteobacteria
Class	Epsilonproteobacteria
Order	Campylobacterales
Family	Helicobacteraceae
Genus	Helicobacter
Species	*H. pylori*

Helicobacter pylori is a Gram-negative, pathogenic bacterium in the human stomach, which can cause severe gastritis, peptic ulcers, and gastric cancer.[1] Depending on the ethnic and geographic origin of humans, the frequency of *H. pylori* occurrence in human individuals varies; however, altogether 50% of the world population is infected with it. *H. pylori* cell is spiral-shaped and polarly flagellated.[2] The motility correlates to infectivity and pathogenicity.[3]

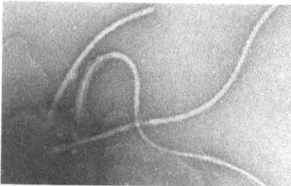

Each filament is enveloped with a membranous sheath, which is physically fragile to preparation and often detaches from the filament. The hook does not look curved, but appears straight (lower).

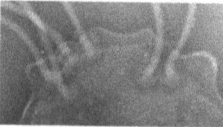

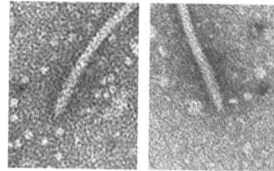

The flagellar base of the *H. pylori* cell shows one of the typical structural features for Gram-negative species (upper). In the isolated basal structure, the C ring structure clearly appears even in the presence of 0.1% Triton X-100 (right).

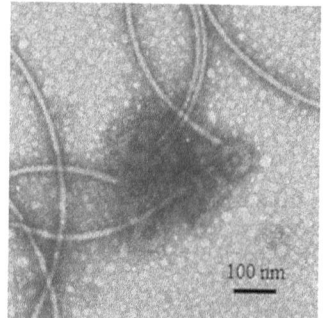

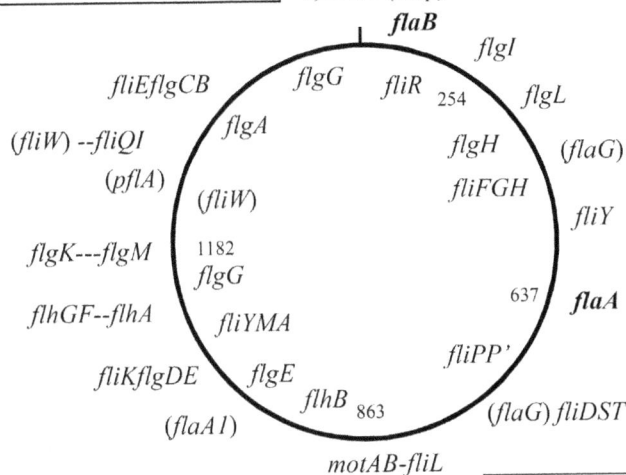

Helicobacter pylori 26695 genome
(1,667,867 bp/ 1,573 genes)

NC_000915
AE000511

Multiple flagellin genes

1,667/0 (Kbp)

Missing genes: *flgF,J, fliJ*

The genome of the *H. pylori* 26695 strain consists of a single 1.6-Mbp chromosome. Flagellar genes are mostly randomly scattered all over the chromosome.[4-6] However, functional units of several geness are maintained: *fliFGH* (C ring formation), *flaG fliDST* (filament cap formation), *fliKflgDE* (hook formation), *fliEflgCB* (rod formation) (see Introduction and ref[5]). Position numbers are omitted for the sake of simplicity.

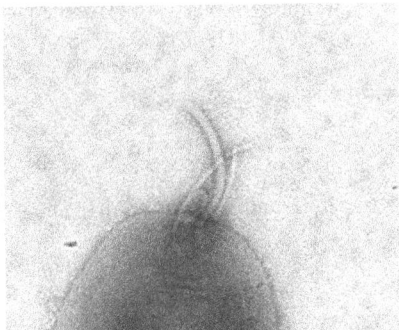

There are two flagellin genes on the chromosome: *flaA* and *flaB* (these flagellins share 58% identical amino acid sequence[4]). Another flagellin candidate *flaA1* is not a true flagellin gene, but is functionally involved in the glycosylation of flagellar proteins.

An *flaA* mutant produces short filaments (left), indicating that FlaA is the major flagellin in the filament and that FlaB is a minor flagellin, which possibly forms the proximal part of the filament.[7]

Strains were provided by Christine Josenhans of Wuerzburg University as of 1995 (present address: Hannover Medical University, Hannover, Germany). The project was carried out by Naohiro Ishikawa.

Gene Arrangement

Arrangement of the flagellar genes differs from species to species. However, there are several groups in arrangement patterns. Taking the flagellar gene arrangement of *Salmonella* as the standard, we can quantify the randomness of the gene arrangement. Suppose the *Salmonella* flagellar gene arrangement is zero:

1. *flgN MABCDEFGHIJKL*
2. *flhEAB cheZYBR-WA motBA flhCD*
3. *fliYZABCDST*
4. *fliEFGHIJKLMNOPQR*

Let's take a gene arrangement (ABCDE). Add +1 every time a gene cluster is cut and separated into two clusters, and add −0.5 when two gene clusters are joined into a new cluster. For example, gene cluster (ADE BC) has two cuts (A/BC/DE) +2 and one join (ADE) −0.5, therefore we get +1.5 for this cluster. Another example: the *Ralstonia* genome, which is closest to the *Salmonella* genome, has 4 cuts and 3 joins, giving it a score of (4 − 1.5 = 2.5). In this way, we obtain scores of all species in this book as follows:

Helpy (ε) 25 > *Magma* (α) 19 > *Gluox* (α) 18 > *Ruegeria* (α), *Azobr* (α) *pof,* 16 > *Braja* (α), *Caucr* (α), *Sinme* (α) 15 > *Azobr* (α) *laf,* 14 > *Rhosp* (α) *fla2,* 12 > *Bdeba* (δ) 11 > *Bacsu* (F), *Selru* (F) 10 > *Entca* (γ), *Paeal* (F) 7.5, *Borbu* (S), 7, *Geoka* (F) 6.5 > *Actmi* (A), 5.5, *Pseae* (γ), 5, *Idilo* (γ) 4.5, *Rhosp* (α) *fla1,* 4 > *Allfi* (γ), 3, *Ralso* (β), 2.5, *Vibpa* (γ), 2 > *Salty* (γ), *Ecoli* (γ), 0 > *Pecca* (γ), −1.

In parentheses, the phylums are indicated: (α) alpha- (β) beta- (β) gamma- (δ) delta- (ε) epsilon-proteobacteria, (F) Firmicutes, (S) Spirochetes.

Flagellar genes of *H. pylori* are the most randomly arranged (+25), whereas those of *Pectobacterium* are least randomly and most compactly arranged (−1) than *Salmonella* genes. It should be noted that this order will be reversed if we take other species, say *B. subtilis*, as the standard strain. However, the most randomly arranged group will stay at the same rank due to the randomness caused by separation of each gene; that is, it will have many cuts, and no rejoins.

Gupta (2000) proposed a hypothesis on gene arrangement: the branching order of different prokaryotic taxa is deduced from the common ancestor as follows: Gram-positives = Archaebacteria> > Spirochetes> > epsilon-, delta- > alpha- > beta- > gamma-proteobacteria. Gupta suggests that the major evolutionary changes within bacteria might have occurred in a directional manner. Our results partly agree with his hypothesis, but there are reverse orders among them. This probably indicates that flagellar genes were laterally transferred as a group independently from the rest of the genes.

Mot Proteins

Mutants that produce paralyzed flagella are called motility deficient (Mot−) mutants. In *S. typhimurium*, there are only 2 *mot* genes (*motA* and *motB*) that are directly involved in motor function. Both MotA and MotB are membrane proteins and they form the stator of the flagellar motor. They associate in a complex with a stoichiometry of MotA(4)MotB(2),[1] and four or five complexes work in a single motor.[2] MotA has a large cytoplasmic domain that interacts with FliG of the C ring to generate torque using the proton motive force (PMF).[3,4] MotB has a large periplasmic domain that interacts with the peptidoglycan layer and stabilizes the motor function.[5]

In *V. alginolyticus*, there are two sets of flagella: polar and lateral. The energy source for each motor is distinguishable; PMF is used for the lateral flagella, whereas the sodium motive force (SMF) is used for the polar flagellum.[6] These unique *mot* genes for the polar flagellum are called *pomA* and *pomB*.[7] If the *flhFG* genes were removed, a polar flagellum would become a peritrichous flagella (see Topic: Flagellar position and shape).

Peritrichously flagellated *B. subtilis* had a second set of Mot proteins, which was SMF-driven and called MotPS.[8–10] This is a somewhat confusing situation; what is the difference between PomAB and MotPS? Another fact makes the situation more complicated. That is, MotAB proteins do not constantly stay at the motor periphery, but are frequently turned over with newly-made motAB proteins.[2] The ion specificity of the motor resides in MotB, PomB, or MotS.[11] In the chimeras MotB-PomB, the ion specificity is exchangeable.[12]

In addition to the conventional *motAB* genes, there are two more *mot* genes (*motX*, and *motY*) in some species. *V. alginolyticus* MotX and MotY are associated with the basal body of the sodium-driven polar flagellum and are required for stator formation together with MotAB.[13–15] *P. aeruginosa* retains one set of *mot* genes (*motA* and *motB*), and a second set (*motC* and *motD*).[16] Although it is indicated that the *motC* and *motD* genes play an important role in pathogenicity, their roles in motor function are not known.

In summary, Mot proteins are not bound at a certain position; rather, they move to where the flagellar motor locates. The ion specificity of Mot proteins is independent from the position where they locate. There are only two sets of Mot proteins essential for motor function: the MotAB complex uses PMF, and the MotPS complex uses SMF.

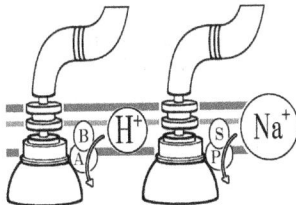

CHAPTER 14

Idiomarina loihiensis — A Habitat of Deep-Sea Volcano

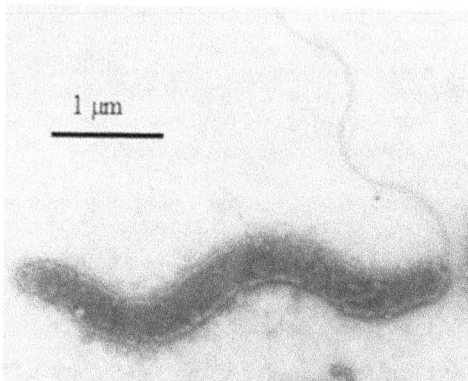

1 μm

Phylum	Proteobacteria
Class	Gammaproteobacteria
Order	Alteromonadales
Family	Idiomarinaceae
Genus	Idiomarina
Species	*I. loihiensis*

Idiomarina loihiensis was isolated from a hydrothermal vent at a 1,300-m depth on the Lōihi submarine volcano in Hawaii. It is a Gram-negative, halophilic bacterium. In contrast to obligate anaerobic vent hyperthermophiles *Thermotoga* spp. and *Pyrococcus* spp., *I. loihiensis* inhabits partially oxygenated cold waters at the periphery of the vent, surviving a wide range of growth temperatures (4°C–46°C) and salinities (from 0.5% to 20% NaCl). The cell contains an abundance of amino acid-transport and degradation enzymes, but neither sugar transport systems nor certain enzymes of sugar metabolism, suggesting that *I. loihiensis* relies primarily on amino acid catabolism for carbon and energy. Each cell possesses a single polar flagellum on a helical body.

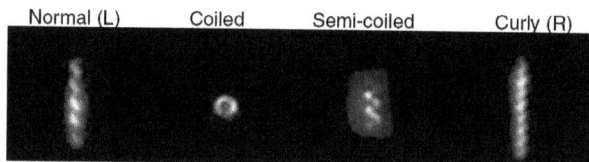

Normal (L)   Coiled   Semi-coiled   Curly (R)

By dark-field microscopy, filaments isolated from *I. loihiensis* show a pair of helical pitch and diameter (1.32 nm, 0.33 nm), which look like *S. typhimurium* Curly filaments (1.29 nm, 0.2 nm). However, *I. loihiensis* filaments are left-handed; the conventional Curly form, on the other hand, is right-handed. Analysis of polymorphs of *I. loihiensis* flagella led us to a new classification of the flagellar family (see Appendix).

Idiomarina loihiensis L2TR genome (2,839,318 bp/ 2,628 genes)	NC_006512 AE017340

2,839/0 (Kbp)

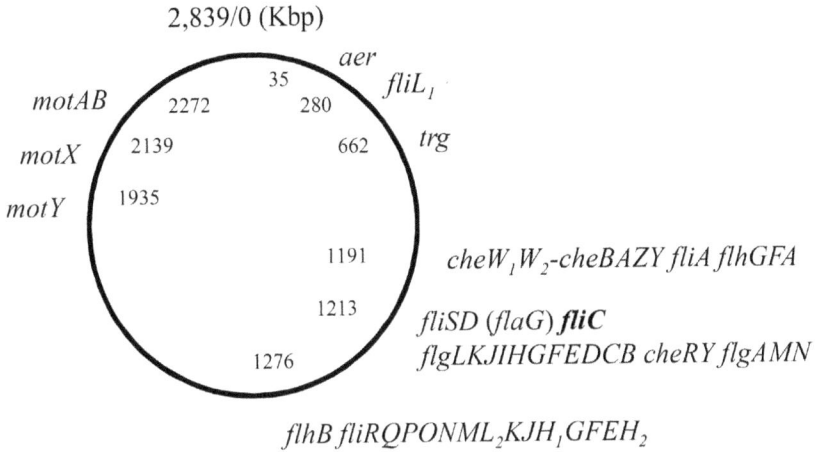

aer

motAB   2272   35   280   *fliL₁*

motX   2139   662   *trg*

motY   1935

1191   *cheW₁W₂-cheBAZY fliA flhGFA*

1213   *fliSD (flaG) **fliC***

1276   *flgLKJIHGFEDCB cheRY flgAMN*

flhB fliRQPONML₂KJH₁GFEH₂

Missing genes: *fliI*

The genome of the *Idiomarina loihiensis* L2TR strain consists of a single 2.8-Mbp chromosome. The majority of the flagellar genes form compact clusters at three loci on the chromosome. The gene arrangement shows characteristics of a polar flagellum with a set of four *mot* genes (*motA*, *motB*, *motX*, and *motY*) and a set of the genes responsible for flagellar number and localization (*flhF* and *flhG*) (see Topic: Mot proteins and Topic: Flagellar position and shape). In the genome, there is only one flagellin gene; the predicted sequence (471 aa) resembles *Pseudomonas aeruginosa* flagellin B (48.3% identity). The purified flagellin runs as a single band at 58 kDa in an SDS-PAGE, and the N-terminal sequence agreed with the predicted one.

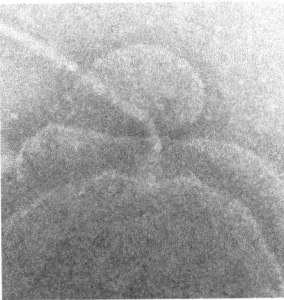

I. loihiensis cells are sensitive to low osmolarity and are easily shocked by washing and staining on the EM grid, consequently giving an image of transparent cells (see Appendix: Protocol IV).

The strains were provided by Maq Alam of University of Hawaii, Honolulu, USA. The project was carried out by Satoshi Shibata.

CHAPTER 15

Legionella pneumophila — Opportunistic Pathogen in Public Bath

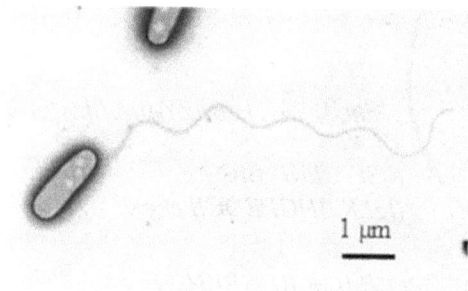

1 μm

Phylum	Proteobacteria
Class	Gammaproteobacteria
Order	Legionellales
Family	Legionellaceae
Genus	Legionella
Species	*L. pneumophila*

Legionella pneumophila is a Gram-negative bacterium that lives in natural and manmade water systems and replicates intracellularly within aquatic protozoa. *L. pneumophila* cells grow aerobically and require L-cysteine-HCl and iron salts for growth. *L. pneumophila* is the only genus of the family.

   L. pneumophila, as the name indicates, causes opportunistic pneumonia in humans. Transmission occurs in the form of aerosolized spray of standing or warm water, and recent occurrences are typified by group infection from air conditioners in gatherings or in the public bath. Infection requires a type IV secretion system called the Dot/Icm system.

   L. pneumophila 130b strain cells produce a single polar flagellum under vigorous aeration, while other strains may possess numerous polar flagella or lateral flagella.

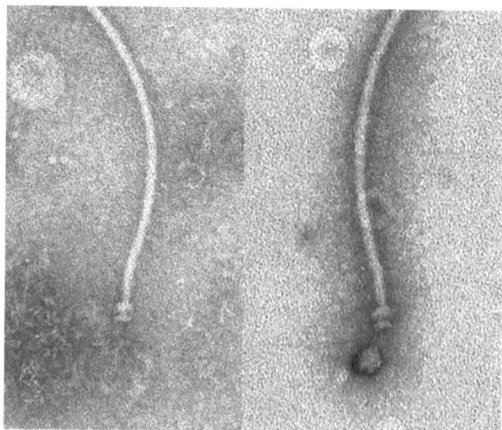

Flagellar filaments purified from *L. pneumophila* cells give a single band at 45 kDa in SDS-PAGE, which agrees with the predicted MW of FliC (471 aa). The outer rings of HBB look similar to those of *V. parahaemolyticus*, but different from those of *Salmonella*.

Legionella pneumophila subsp. *pneumophila* Philadelphia 1 genome (3,397,754 bp/ 2,943 genes)

NC_002942

3,397/0 (Kbp)

dot/icm

flhB'

icm

3021

2912

479

517   (*rpoN*)

929   (*fleQ*)

motAB   2619

981

motBC fliA flhGFAB
fliRQPONM

1988

flgNMA

1961   1342

flgBCDEFGHIJKL

1879   1475

fliJIHGFE (fleR) (flrB)   *fliK*

fliSD (flaG) fliC

Missing gens: *fliL*

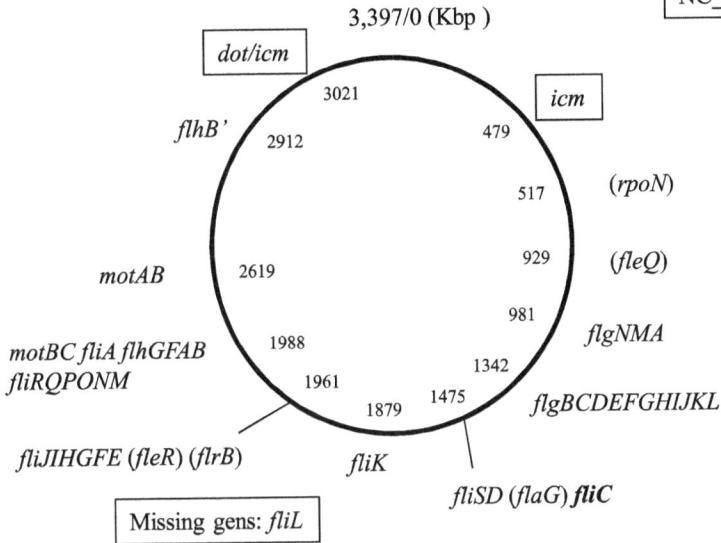

The genome of *Legionella pneumophila* Philadelphia 1 strain consists of a single 3.5-Mbp chromosome. The flagellar gene regulation cascade in *L. pneumophila* is similar to that described in *P. aeruginosa*, in which *fleQ* and *rpoN* are involved in flagellar gene regulation (see Topic: Gene regulation). There are two sets of *motABIC* genes. From the sequence analysis, *motBC* belongs to the proton-driven motor, while *motAB* belongs to the sodium-driven motor (see Topic: Mot proteins). The type IV secretion system is composed of *dot/icm* genes, whose clusters are shown in the box. The *fleR* gene encodes a sigma54-dependent response regulator, and *flrB* encodes a sensor kinase (see Chapter 2: *A. fischeri*).

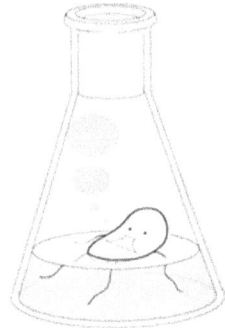

Strains were provided by Hiroki Nagai and Tomoko Kubori of Osaka University, Osaka, Japan.
The project was carried out by Saori Higaki and Aska Kataoka.

CHAPTER 16

Magnetospirillum magnetotacticum — High-Quality Magnet in the Pond

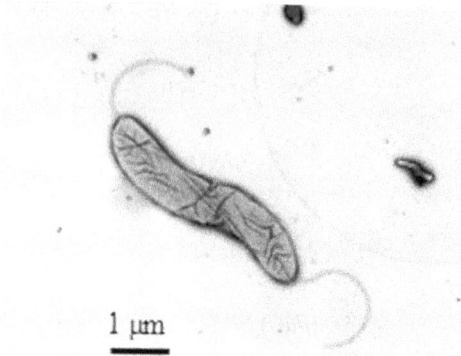

1 μm

Phylum	Proteobacteria
Class	Alphaproteobacteria
Order	Rhodospirillales
Family	Rhodospirillaceae
Genus	Magnetospirillum
Species	*M. magnetotacticum*

Magnetospirillum (formerly *Aquaspirillum*) *magnetotacticum* is a Gram-negative, helical (clockwise), magnetotactic, microaerophilic spirillum. It can be isolated from areas associated with the sediment–water interface of many freshwater environments. The *M. magnetotacticum* cell synthesizes high quality single-domain magnetite crystals (dots on a straight line in the middle of the cell body), which are far superior to those produced industrially. Each cell possesses a single flagellum at both poles, a good representative of the bipolar flagella.

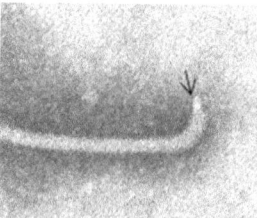

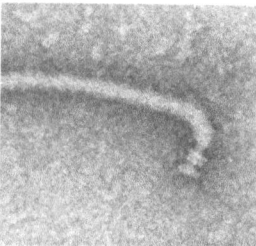

The hook of *M. magnetotacticum* cell appears thicker than the filament and the rod (arrow), probably due to the large molecular weight of FlgE. The hook length is shorter than is inferred by the large molecular size of FliK (587 aa) (see Topic: Hook length).

The basal body looks the same as that of *Salmonella*. Again, the hook looks thicker than the PL ring of the basal body.

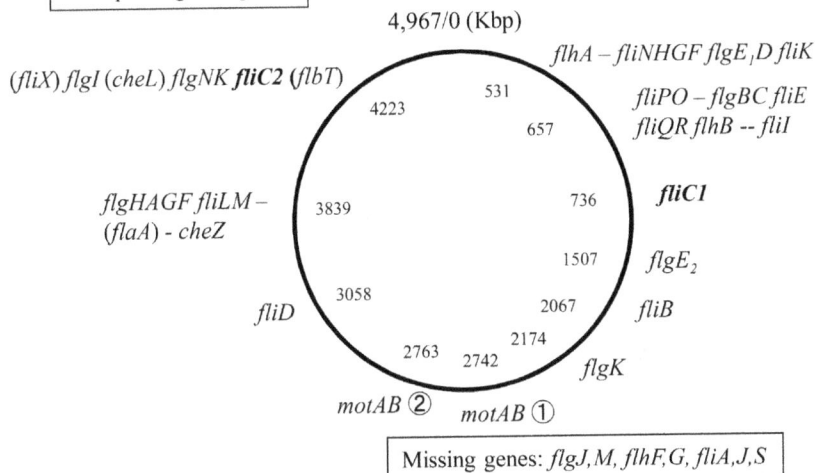

The genome of the *Magnetospirillum magneticum* AMB-1 strain consists of a single 4.9-Mbp chromosome. Flagellar genes are scattered at several sites on the chromosome. The gene arrangement is similar with those of *C. crescentus* and *B. japonicum*. There are two flagellins: FliC1 (489 aa) and FliC2 (712 aa), which is one of **the large flagellins** (see Topic: Flagellin size). However, an isolated filament sample gave rise to a single band at 36 kDa in SDS-PAGE. Therefore, FliC1 is the major flagellin, while FliC2 plays, if any, a minor role in filament assembly. There are two *flgE* genes: $FlgE_1$ (843 aa) and $FlgE_2$ (545 aa), which are also much larger than the ordinary FlgE. There are also two sets of MotAB proteins. Physiological experiments using ionophores indicate that both MotAB proteins are for a proton-driven motor (see Topic: Mot proteins).

Each cell possesses bipolar flagella. However, the *flhFG* genes that are necessary for placing the flagellum at a pole have not been found. The *M. magnetotacticum* cell has the flagellin-specific methylase, FliB, which exists in a few species such as *Salmonella* and *H. pylori*.

Strains were provided by Yoshihiro Fukumori of Kanazawa University, Kanazawa, Japan.
The project was carried out by Kaoru Uchida.

Flagellin size

There are large differences in molecular size among flagellins of various species. The basic architecture of flagellin consists of 7 structural domains in the N- and C-terminal regions. The molecule is also divided by homology conservation into two regions: a conserved (C) region and a variable (V) region.

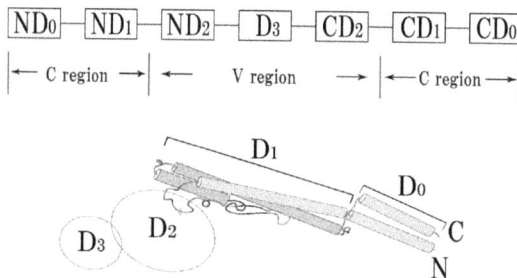

The conserved region includes both termini of ca.100 aa length and forms alpha-helix bundles in a filament (D_0 and D_1 domains). New flagellin genes are identified by using the high homology in the conserved region. The variable region occupies the central domains (D_2 and D_3), which actually extend out into the three-dimensional structure of the filament.[1] This extended part can be deleted without affecting the filament assembly.[2]

A part of the extended structure (D_3) forms a hyper-variable domain, which becomes an epitope that stimulates the acquired immune system as an H antigen and subsequently creates a variety of serotypes of *Salmonella* strains.[3] Flagellin is also recognized by the toll-like receptor 5 (TLR5) in the innate immune system of humans and animals.[4] The epitope recognized by TLR5 locates in the N-terminal conserved region between ND_1 and ND_2.[5]

Enteric bacteria tend to have flagellin with a large molecular size, plausibly to escape attack by the human immune system. On the contrary, environmental species that live in the soil, fresh water, or sea water have flagellins of small molecular size. The smallest flagellin of *Ralstonia* lacks most of the D_2D_3 domains, leaving minimal functional domains (D_0D_1): filament polymerization and holding helicity. The large flagellins in *Gluconococcus* and *Magnetobacteria* are not the major component of the filament, but a minor one with unknown function.

Table T4.1 Flagellin Sizes of the Species Appearing in this Book

Large flagellins (aa)		Medium flagellins (aa)		Small flagellins (aa)	
G.. oxydans FliC2	792	E. coli	497	C. crescentus	297
M. magnetotacticum FliC2	712	R. sphaeroides Flag-1	493	G. kaustophilus Laf	297
S. ruminantium	699	G. oxydans FliC1	491	P. carotovorum	288
G. kaustophilus Pof	604	Salmonella	489	Ruegeria sp.	282
A. brasilensis Pof	621	M. magnetotacticum FliC1	489	R. sphaeroides Flag-2	281
S. degradans	587	L. pneumophila	475	B. japonicum Laf	274
P. alvei	576	I. loichiensis	471	B. bacteriovorus	274
S. flexneri*	550	A. missouriensis	411	A. brasilensis Laf	274
H. pylori	509	X. oryzae	399	R. solanacearum	273
		S. meliloti	395	M. voltae**	231
		P. aeruginosa	393		
		V. parahaemolitycus	376		
		E. casseliflavus	358		
		B. burgdorferi	336		
		B. japonicum Pof	314		
		B. subtilis	304		

*S. flexneri *does not express the flagellin gene.*
**M. Voltae *is an Archaeal bacterium and may have a flagellar structure different from the eubacterial flagellum (see Chapter 32:* M. voltae).

Flagellin as a tool for biotechnology

Flagellin molecules can be used as a biotechnological tool for two reasons. First, flagellins are secreted in considerable amounts into the culture medium without cleavage of the N-terminal region. Second, the fairly large portion of flagellin can be deleted. Thus, it is to be expected that medically important peptides with an appropriate sequence can be inserted into the deleted part of the central variable region and overproduced as a pseudo-flagellin. An earlier pioneering attempt to overproduce peptides as large as the central region had been unsuccessful, but it had been limited to the small peptides.[6] Since then, many attempts to insert larger peptides have been successfully undertaken. Now, large fragments of bacterial adhesins up to 302 amino acids have been functionally inserted.[7] Various oligopeptides have been successfully displayed on the surface of flagellar filaments.[8–10] Using FlgM, another flagellar secretion substrate, as a carrier protein, various neuroactive peptides and proteins from snails, spiders, snakes, sea anemone, and bacteria have been secreted via the bacterial flagellar T3SS and successfully recovered from the culture medium by one step purification.[11]

CHAPTER 17

Paenibacillus alvei — Flagella-Dependent Social Motility

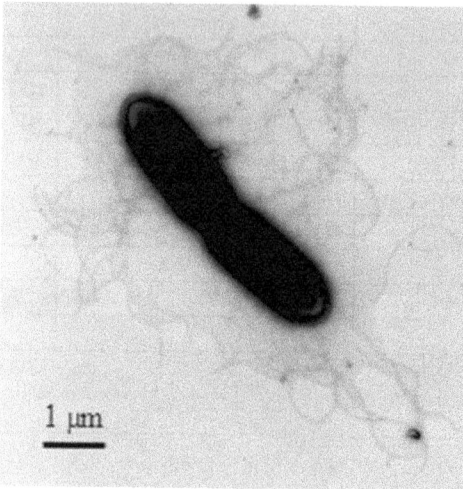

1 μm

Phylum	Firmicutes
Class	Bacilli
Order	Bacillales
Family	Paenibacillaceae
Genus	Paenibacillus
Species	*P. alvei*

The *Paenibacillus* species are Gram-positive, rod-shaped aerobes. Some of them exhibit **cooperative motility**, resembling "S motility" by *Myxococcus xanthus* (see Chapter 33), and show several morphotypes on agar plates. *P. alvei* strains clinically isolated from a dog mouth are notable for their motile microcolonies that migrate over an agar surface to form **nebula-like colonies** (below left).

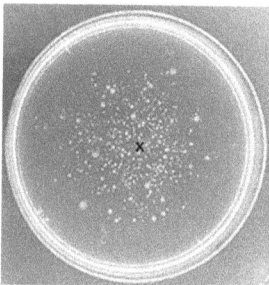

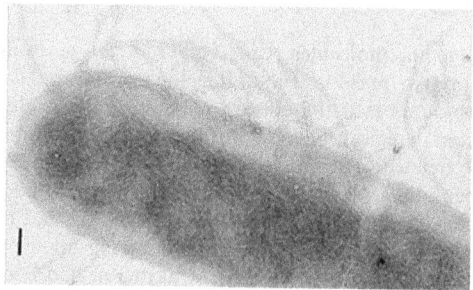

Cells were inoculated at the center (x) of a hard (1.5%) agar plate and incubated at 37°C overnight. A nebula pattern is formed on the plate. Pattern formation is considerably affected by concentrations of salt and agar, and by temperature.

When grown on agar plates, the cells produce numerous peritrichous flagella, but no flagella are formed in liquid culture. Under aerobic conditions, cells actively swarm over surfaces. The flagellar filament is composed of a single flagellin with apparent molecular weight of 30 kDa by SDS-PAGE. Isolated filaments melt at pH 3.0, leaving intact HBB.

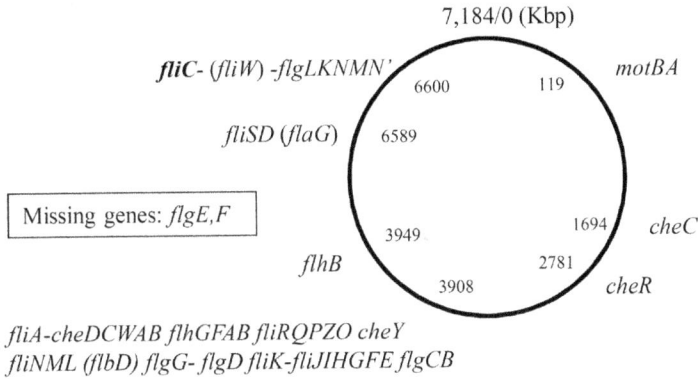

| *Paenibacillus* sp. JDR-2 genome (7,184,930 bp/ 6,213 genes) | NC_012914 PRJNA20399 CP001656 |

7,184/0 (Kbp)

fliC- (*fliW*) -*flgLKNMN'*

fliSD (*flaG*)

Missing genes: *flgE,F*

flhB

motBA

cheC

cheR

6600   119
6589
3949   1694
3908   2781

fliA-cheDCWAB flhGFAB fliRQPZO cheY
fliNML (*flbD*) *flgG*- *flgD fliK*-*fliJIHGFE flgCB*

The genome of *Paenibacillus* sp. JDR-2 strain consists of a single 7.1-Mbp chromosome. The majority of its flagellar genes form a compact cluster at around 3,908 Kbp. The gene arrangement is the same, but in the opposite direction, as that of *S. ruminantium* (see Chapter 25). There is one flagellin gene. FliC (576 aa) is a little larger than the observed MW (30 kDa) of a single band by SDS-PAGE.

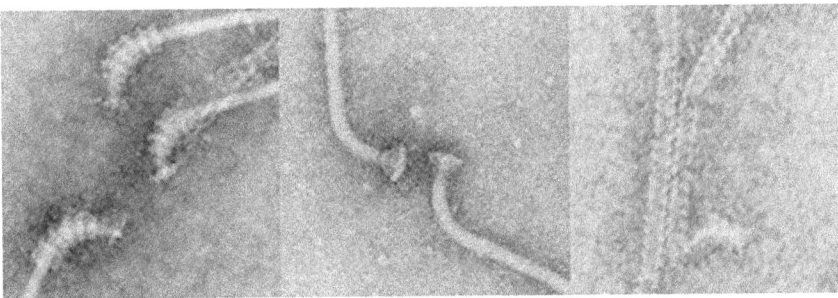

The hook region is covered by a **helical filament** (left), which is a new structure and tentatively named the hook cover. The hook cover can be slipped off (middle) and forms a longer filament by end-to-end aggregation (right). The pitch and diameter of the helix is 12.8 ± 1.6 nm and 28.4 ± 1.6 nm. A helix has 6 pitches, which is just long enough to cover the hook length of 63.6 ± 6.1 nm. The function of the hook cover is not known.

Strains were provided by Ryo Harasawa of Iwate University, Morioka, Japan. The project was carried out by Yoshika Nosaka, and Kyohei Miyauchi.

CHAPTER 18

Pectobacterium carotovorum — Subpolar Hyper-Flagellation

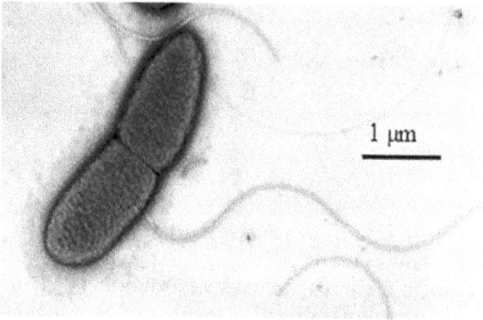

1 μm

Phylum	Proteobacteria
Class	Gammaproteobacteria
Order	Enterobacteriales
Family	Enterobacteriaceae
Genus	Pectobacterium
Species	*P. carotovorum*

Pectobacterium (formerly *Erwinia*) *carotovorum* is a Gram-negative plant-specific pathogen, causing soft rot disease of various plant hosts, and blackleg in potato by degradation of the plant cell wall. *P. carotovorum* cells colonize the intercellular spaces of plant cells and deliver potent effector molecules (Avr – avirulence) through a type III secretion system (see Topic: Flagella and pathogenicity). Various exoenzymes are secreted via a type II secretion system that degrade the plant cell wall by depolymerization of the pectin component.

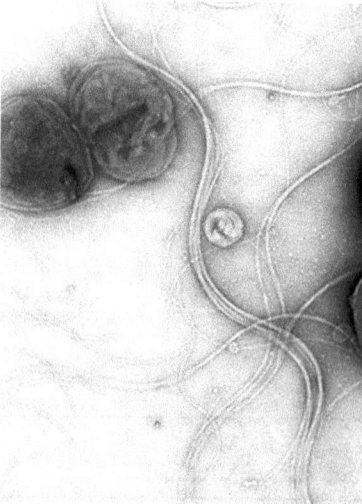

The *P. carotovorum* EC1 strain possesses a single **subpolar flagellum** when grown in an ordinary nutrient medium at 30°C or higher (upper panel). In contrast, the cells become hyperflagellated and highly motile when grown in a medium containing fructose at 26°C. Filaments tend to aggregate to form thick and **long bundles of filaments** (left).

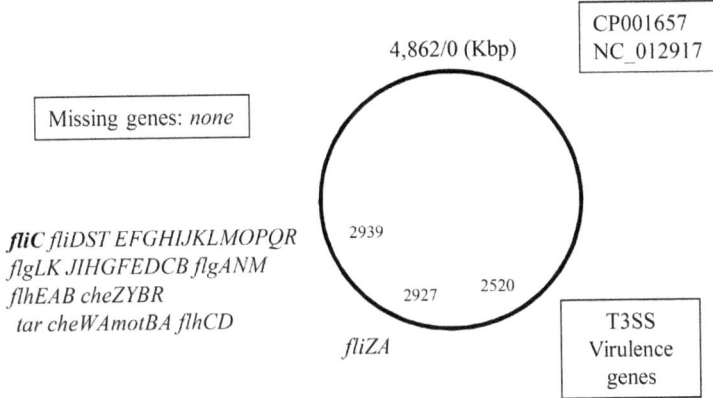

**Pectobacterium carotovorum subsp. carotovorum PC1 genome
(4,862,913 bp/ 4,246 genes)**

4,862/0 (Kbp)

CP001657
NC_012917

Missing genes: *none*

*fliC fliDST EFGHIJKLMOPQR
flgLK JIHGFEDCB flgANM
flhEAB cheZYBR
tar cheWAmotBA flhCD*

2939

2927   2520

fliZA

T3SS
Virulence
genes

The genome of *Pectobacterium carotovorum* PC1 strain consists of a single 4.8-Mbp chromosome. As a member of the Enterobacteriaceae, *P. carotovorum* is related to *Ecoli/Salmonella*. Accordingly, the flagellar genes are clustered as orderly as, and more compactly than, that of *Ecoli/Salmonella*. This is **the most densely packed cluster of flagellar genes** among all of the species in this book (see Topic: Gene arrangement). It is noted that the genome retains the *fliZ* gene, which is necessary for expression of pathogenicity in *S. typhimurium*. Pathogenicity Island including the *hrp*, *hrc* genes is indicated in a box.

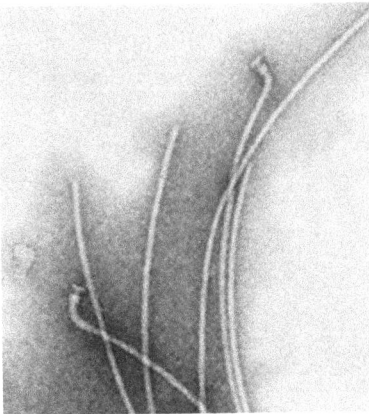

There is only **one flagellin gene** *fliC* on the chromosome. Purified filaments give rise to one band at 31 kDa by SDS-PAGE. The amino acid sequences of the terminal regions of *P. carotovorum* FliC (288 aa) show a high identity (more than 70 %) with those of *Salmonella* FliC (489 aa).

Strains were provided by Shinji Tsuyumu of Shizuoka University, Shizuoka, Japan. The project was carried out by Satoshi Shibata.

CHAPTER 19

Pseudomonas aeruginosa — Opportunistic Pathogen in the Hospital

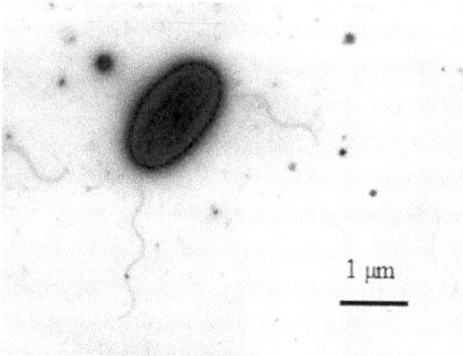

1 μm

Phylum	Proteobacteria
Class	Gammaproteobacteria
Order	Pseudomonadales
Family	Pseudomonadaceae
Genus	Pseudomonas
Species	*P. aeruginosa*

Pseudomonas aeruginosa is a Gram-negative, opportunistic human pathogen.[1] *P. aeruginosa* is a very ubiquitous microorganism, for it has been found in environments such as soil, water, humans, animals, plants, sewage, and hospitals.[2]

Each cell possesses both flagella and pili. They swim by means of polar flagellum in liquid. Its strains have either a-type or b-type flagella, by a classification that is based on the antigenicity of the flagellin subunit.

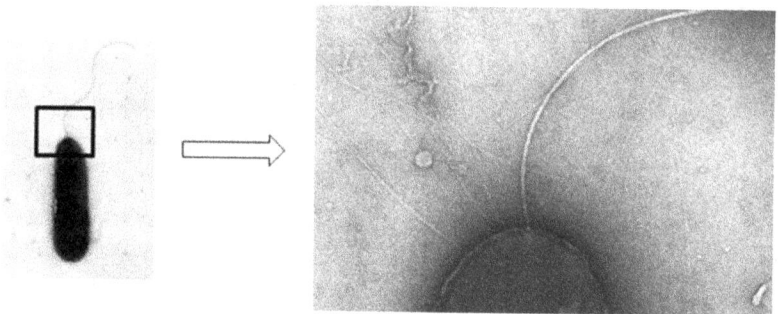

P. aeruginosa pili contribute greatly to its ability to adhere to mucosal surfaces and epithelial cells.

Pseudomonas aeruginosa PAO1 genome (6,264,404 bp/ 5,571 genes)	NC_002516 AE004091

6,264/0 (Kbp)

flgBCDEFGHIJKL (fgtA)
fliC (flaG) fliDS (flePQ)
(fleSR) fliEFGHIJ

motAB   5557

1162

1570

fliKLMNOPQR flhB - flhAFG
fliA cheYZAB (motCD) cheW

4057

(rpoS)

3760

Missing genes: *none*

cheRW flgAMN

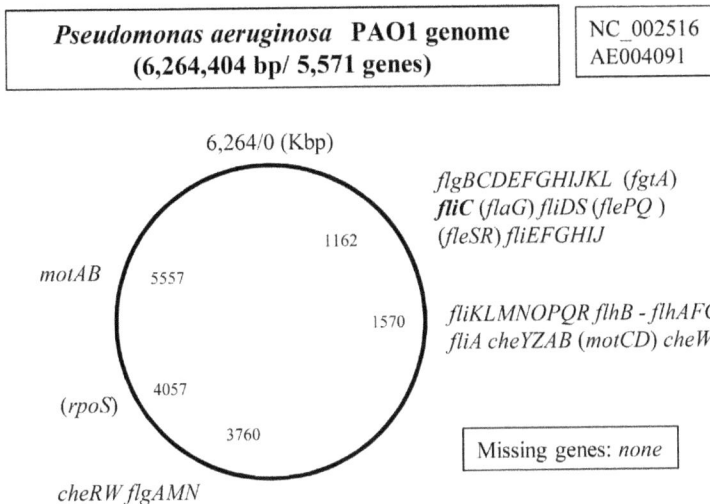

The genome of the *Pseudomonas aeruginosa* PAO1 strain consists of one 6.2-Mbp chromosome and numerous chromosome-mobilizing plasmids that are very significant to the organism's lifestyle as a pathogen. The majority of the flagellar genes form compact clusters at two loci on the chromosome. The gene arrangement is unique and thus makes its own paradigm, which is different from that of *Salmonella* (see Topic: Gene regulation). The *fgtA* gene is involved in glycosylation of flagellin.[3]

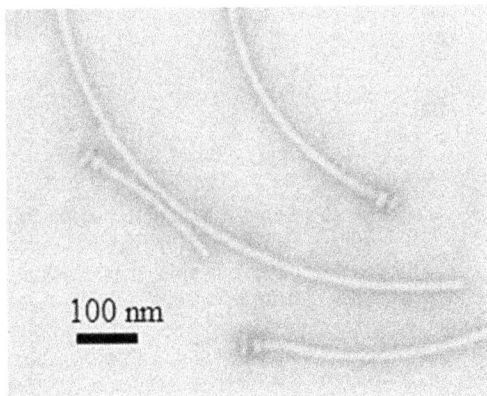

100 nm

The *P. aeruginosa* PAO1 strain carries b-type flagella, while the PAK strain carries a-type flagella. PAO1 FliC (488 aa) is exchangeable with PAK FliC (394 aa) to form filaments *in vivo* and *in vitro*. However, they are not exchangeable with *Salmonella* flagellin.[4] The difference in MW between the two flagellins comes from the different sizes of the central region, which determines the serotypes.

Strains were provided by Reuben Ramphal of University of Florida, Gainesville, USA.

Twitching motility & pathogenicity

Pseudomonas aeruginosa PAO1 genome

6,264/0 (Kbp)

algRZH

exoT

5922

587

pilABCD- - - pilSR- -
fimTU pilVWXY1Y2E

5069

831

algU

pilGHIJK chpABCDE

4496

pilTU

4365

1840

pscUTSRQPON
popNpcr1234
pcrDRGVHpopBD exsC-
BAD pscBCDEFGHIJKL

4303

exoS

4264

2410

pilF

T3SS

exoY

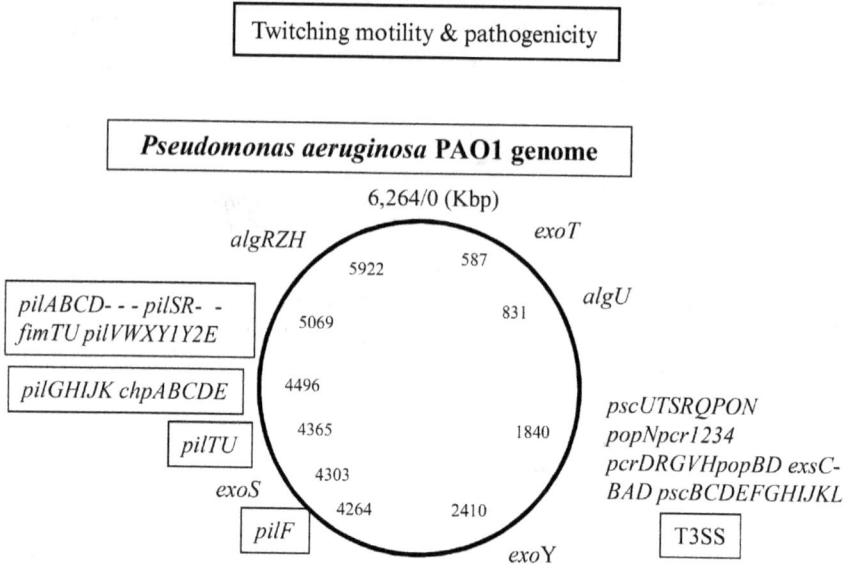

The *pil* genes (in box) are required for assembly of **the type IV pilus** that is used for twitching motility,[5] the flagella-independent surface motility,[6] as seen in *M. xanthus* (see Chapter 33). The type IV pili and alginate are two important virulence determinants. AlgR-FimS is a two-component signal transduction system.[7] Alginate production and twitching motility are closely related. The *psc*, *pcr*, *pop*, and *exo* genes are required for constructing **the type III (virulence) secretion system** (T3SS)[8] (see Topic: Flagella and pathogenicity).

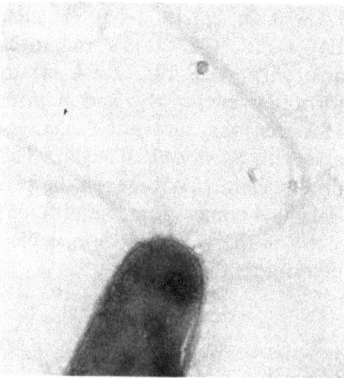

P. aeruginosa strains produce numerous pili on the cell surface (left), which form bundles of pili. The pili's tip is responsible for the adherence to the host cell surface. The type IV pili are long polar filaments composed of homopolymers of pilin, PilA. Flagellar filament is also involved in pathogenicity by adhesion.[9]

Flagella and Pathogenicity

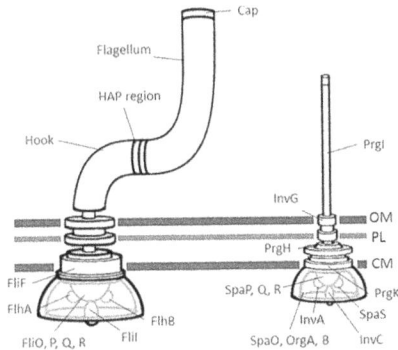

It is known that flagellar motility is involved in the expression of pathogenicity in both plant and animal pathogens.[1-8] In *S. typhimurium*, two flagellar genes, *fliT* and *fliZ*, regulate the gene expression of Salmonella Pathogenicity Island 1 (SPI1)[9,10] which encodes components of type III secretion system (T3SS) (see Chapter 24). The needle complex, the major structure of the secretion apparatus, was first isolated from *S. typhimurium*.[11]

The structure of the needle complex resembles the basal structures of the flagellum[12] (above), and their component proteins have high sequence homology (below). Three proteins (FliPQR) that form the secretion gate are especially highly homologous through species, indicating that both structures might be derived from a common ancestral origin.[13]

Table T5.1 Major Component Proteins for the Type III Secretion Apparatus								
Species	1	2	3	4	5	6	7	8
SPI-1*	SpaP	SpaQ	SpaR	SpaS	InvA	InvC	PrgK	PrgI
SPI-2*	SsaR	SsaS	SsaT	SsaU	SsaV	SsaN	SsaJ	SsaG
Yersinia	YscR	YscS	YscT	YscU	LcrD	YscN	YscJ	YscF
Shigella	Spa24	Spa9	Spa29	Spa40	MxiA	Spa47	MixJ	MxiH
EPEC**	EscR	EscS	EscT	EscU	EscV	EscN	EscJ	EscF
Pseudomonas	PscR	PscS	PscT	PscU	PscD	PscN	PscJ	PscF
Ralstonia	HrcR	HrcS	HrcT	HrcU	HrcV	HrcN	HrcJ	HrpX***
Flagella	FliP	FliQ	FliR	FlhB	FlhA	FliI	FliF	FlgE****
Function	Secretion gate			Secretion gate keeper		ATPase	Base	Needle/hook

*SPI-1 and SPI-2 are the two Salmonella pathogenicity islands (see Chapter 24).
**Enteropathogenic Escherichia coli (see Chapter 10).
***The needle structure in Ralstonia species has not been identified. The filamentous part of the apparatus is called Hrp pili, which is composed of HrpY.
****The hook is structurally not a counterpart of the needle. However, lengths of both structures are controlled by a soluble and secretable protein, FliK and the homologs, by a similar mechanism (see Topic: Hook length).

CHAPTER 20

Ralstonia solanacearum — Ubiquitous Plant Pathogen

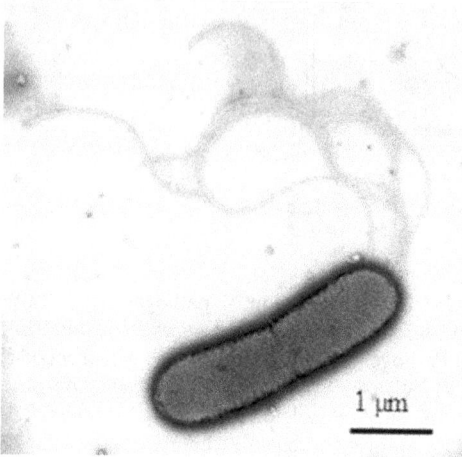

Phylum	Proteobacteria
Class	Betaproteobacteria
Order	Burkholderiales
Family	Alcaligenaceae
Genus	Ralstonia
Species	*R. solanacearum*

Ralstonia solanacearum is a Gram-negative, ubiquitous plant pathogen, which is responsible for infection of over 200 plants species.[1] *R. solanacearum* cells secrete a copious amount of extracellular polysaccharide into the stem vessels to cause the death of the plant by wilt disease.[2] Each cell possesses several polar flagella. Motility by flagella allows cells to invade their host.[3] Once inside the host, the cells lose flagella and colonize by type IV pili. *R. solanacearum* needs aerotaxis for biofilm formation.[4] *R. solanacearum* is the **only example of betaproteobacteria** in this book.

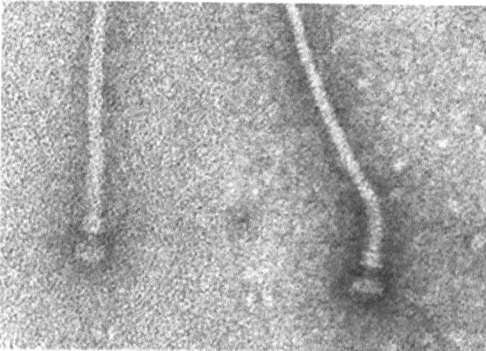

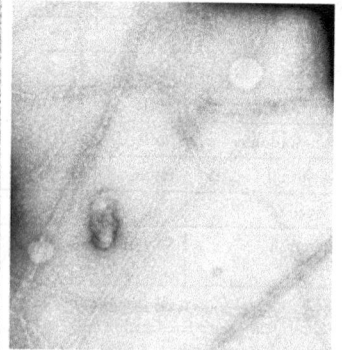

R. solanacearum cells produce flagella (left) that resemble *Salmonella* flagella and pili (right) for adhesion.

Ralstonia solanacearum genome (5,810,922 bp/ 5,113 genes)	NC_003296

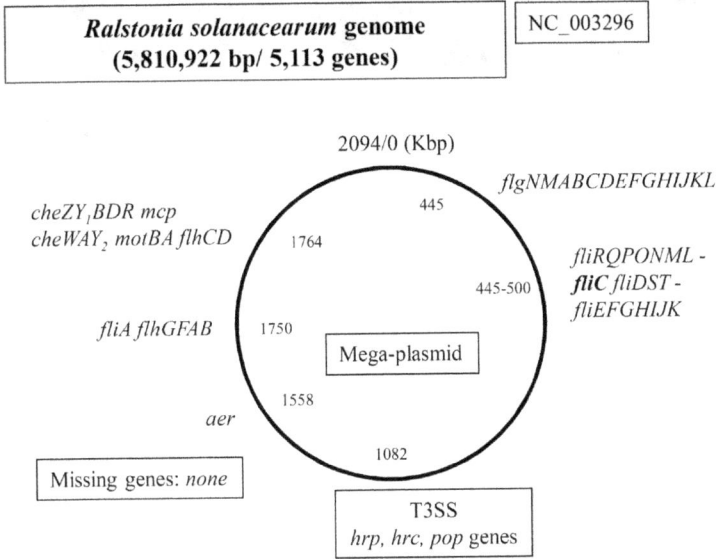

The genome of the *R. solanacearum* strain consists of two replicons: one 3,716,413 bp chromosome and one 2,094,509 bp mega-plasmid.[5] Although *R. solanacearum* is phylogenetically far from *Ecoli/Salmonella*, their flagellar systems are surprisingly similar to each other. The flagellar genes form several **compact clusters on the megaplasmid**. The gene arrangement and regulation resemble those of *Salmonella*; the flagellar gene expression is regulated by the master genes (*flhCD*) and sigma 28 (*fliA*), together with anti-sigma factor (*flgM*).[6] There is one *fliC* gene (273 aa), which is one of **the smallest flagellins** (see Topic: Flagellin size).[7]

Some of the Type III virulence genes (*hrc, hrp, pop*) are also located at one locus on the megaplasmid.[8,9] Genes required for assembly of type IV pilus are located on the chromosome (not shown).

Stains were provided by Takafumi Mukaihara Research Institute for Biological Sciences, Okayama, Japan.
The project was carried out by Shin-no-suke Nakamura and Kaoru Uchida.

CHAPTER 21

Rhodobacter sphaeroides — A Resourceful Little Bug

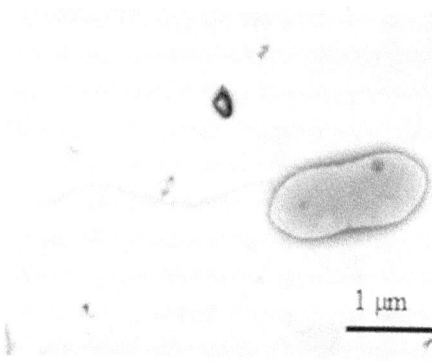

Phylum	Proteobacteria
Class	Alphaproteobacteria
Order	Rhodobacterales
Family	Rhodobacteraceae
Genus	Rhodobacter
Species	*R. sphaeroides*

Rhodobacter sphaeroides is a Gram-negative, facultative photosynthetic bacterium. *R. sphaeroides* cells can live both in fresh water and sea water, and form a pinkish film on the surface of ponds. In addition to photosynthesis, *R. sphaeroides* shows a wide range of metabolic capabilities which include lithotrophy, aerobic and anaerobic respiration, nitrogen-fixation, and synthesis of tetrapyrroles, chlorophylls, heme, and vitamin B12. Many strains of *R. sphaeroides* possess one flagellum located on the side of the cell body, but the flagellum is actually peritrichous (see Topic: Flagellar position and shape).

R. sphaeroides WS8N strain has two flagella systems: Fla1 and Fla2. Fla1 produces a single subpolar flagellum like that of the other strains, while Fla2 produces polar flagella. Fla2 was discovered in 2011, and analysis of its flagella is under way.

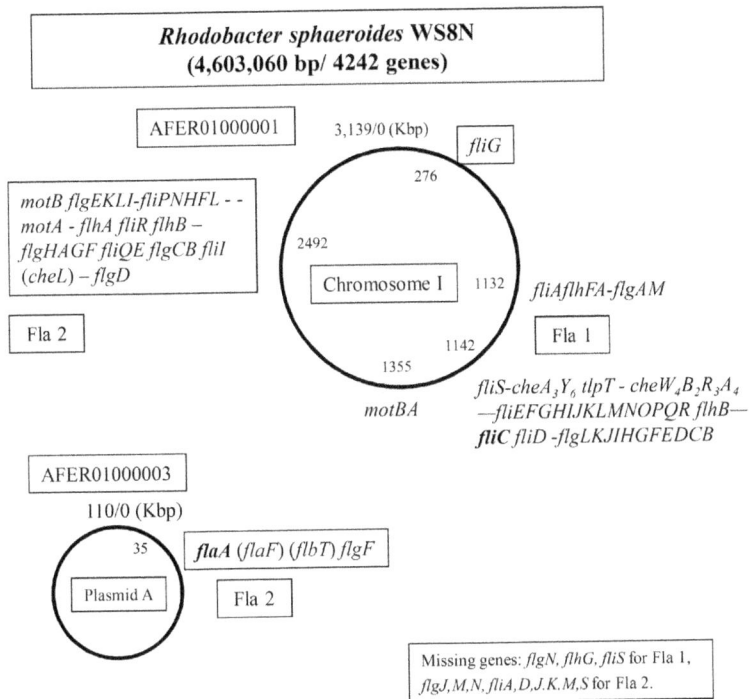

The genome of the *R. sphaeroides* WS8N strain consists of **two chromosomes** (I: 3139278 bp and II: 968108 bp) **and five plasmids** (A–E). The majority of the flagellar genes for Fla 1 and Fla 2 (in box) are clustered at separate loci on chromosome I (3.01-Mbp) and on plasmid A (0.1-Mbp). The organization of Fla 1 shows similarities to that seen in the entire single set of flagellar genes of *E. coli/Salmonella*, while that of Fla 2 shows a random arrangement of genes, which is typically seen in *Bradyrhizobium japonicum* and *Helicobacter pylori*. Most flagellar genes for the Fla 1 system have already been found, while many of those for the Fla 2 system are missing and are yet to be found.

R. sphaeroides WS8N strain was provided by Georges Dreyfus of UNAM, Mexico City, Mexico.

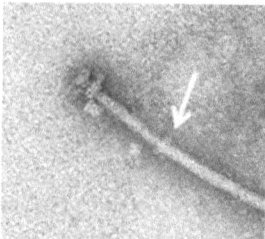

The HAP region (arrow) appears extruded and larger than that of *Salmonella* species, probably due to the large FlgK (1,363 aa). This is one of **the largest FlgK** so far reported. Compare with *Salmonella* FlgK (552 aa).

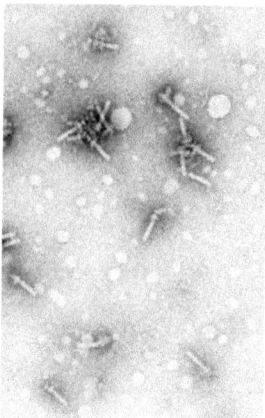

The hook of the polar flagellum characteristically appears **straight** between pH 4 and 9. The hook length of *R. sphaeroides* is 71.0 ± 6.7 nm, which is one of **the longest hooks** in nature so far studied (see Topic: Hook length).

The Fla1 flagellar filament is composed of a 59 kDa flagellin, whose sequence corresponds to that of the sole copy of *R. sphaeroides fliC* (493 aa). The hook protein runs as a 44 kDa band and the sequence of the N-terminus corresponds to that of FlgE (423 aa).

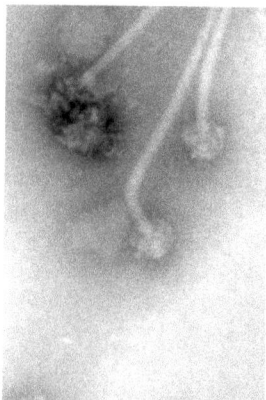

The *R. sphaeroides* HBB is acid labile, but heat stable. Crude samples of intact flagella often retain the C ring still attached to the basal structure.

Strains were provided by Liz Sockett of Nottingham University, Nottingham, U.K.

Flagellar Position and Shape

The position of a flagellum growing on a cell is classified into four groups: polar (at one or both poles), subpolar (near the pole), lateral (from the middle half of the cell body), and peritrichous (randomly arranged on the cell body).[1] Lateral flagella are rare in nature and there are only two examples: *Selenomonas ruminantium* and *Lachnospira multiparus*, both of which are anaerobes inhabiting the cow rumen (see Chapter 25).[1,2]

Today, the term "lateral" is often confused with the term "peritrichous." Two types of flagella in *Vibrio parahaemolyticus* are called the polar flagellum and the lateral flagella (see Chapter 27).[3,4] However, the lateral flagella are not distinguishable from peritrichous flagella. *Rhodobacter sphaeroides* cells possess one flagellum located on the side of the cell body. The flagellum has been called a lateral or subpolar flagellum.[5] Statistical analysis of the position of a single flagella revealed that it is placed evenly between the pole and the midpoint of the cell body,[2] indicating that it is actually peritrichous. There is a reason for this confusion. The pitch of lateral flagella is much smaller than that of peritrichous flagella, which would tend to indicate that a flagellar shape is defined by its growing position. There is a line of evidence against this belief, however. Two different types of flagella in *B. japonicum* (see Chapter 7) grow at the subpole. In *flhFG* (genes required for placing flagella at a pole)[6-8] mutants, flagella move from a pole to the peritrichous position without changing shape.[9] Therefore, flagellar position and flagellar shape can be independent from each other. The apparent appearance of *B. subtilis* flagella is peritrichous, but it was recently found that they are arranged in a more orderly fashion,[10] suggesting that even the peritrichous position may not be randomly arranged, but precisely planned to a certain position.

Flagellar shape is defined by the helix parameters: helix pitch and diameter. We classified flagellar shape into four groups: Family I (large pitch), Family II (medium pitch), Family III (small pitch), and Family IV (exceptions: irregular pitch)[11] (see Appendix). In theory, there are 16 possible combinations of flagellar position and shapes; 4 (polar, subpolar, lateral, peritrichous) × 4 (Family I, Family II, Family III, Family IV) = 16. In reality, only about half of them are found in nature.

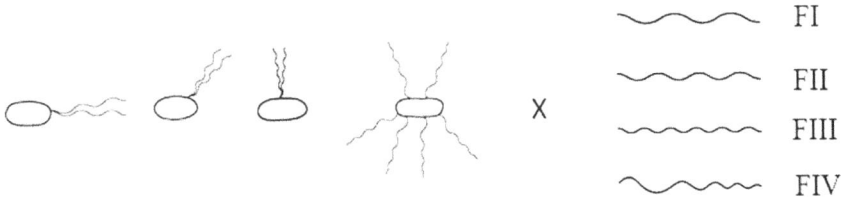

CHAPTER 22

Ruegeria sp. TM1040 — A Fast Swimmer in the Sea

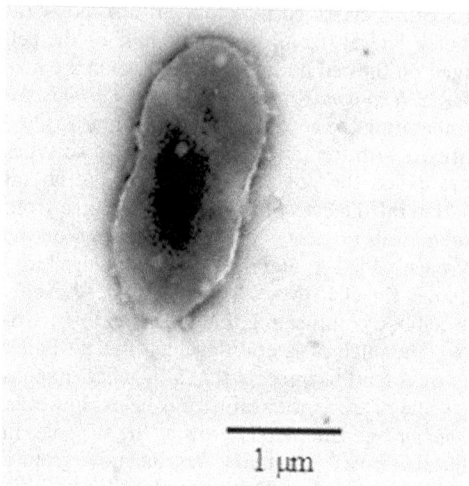

Phylum	Proteobacteria
Class	Alphaproteobacteria
Order	Rhodobacterales
Family	Rhodobacteraceae
Genus	Ruegeria
Species	*Rugeria Sp.*

1 μm

Ruegeria (formerly *Silicibacter*) sp. TM1040 is a Gram-negative ovoid or rod-shaped bacterium and a member of the marine Roseobacter that forms symbioses with unicellular eukaryotic phytoplankton, such as dinoflagellates. The symbiosis is complex and involves a series of steps that physiologically change highly motile bacteria into cells that readily form biofilms on the surface of the host by the biphasic **"swim-or-stick" switch**. Cells lacking wild-type motility fail to establish biofilms on host cells and do not produce effective symbioses. Each cell possesses one or a few polar flagella and can **swim as fast as ~100 μm/s**, which is one of the fastest swimming bacteria in the sea. Compare with the fastest swimmer in fresh water (Chapter 1).

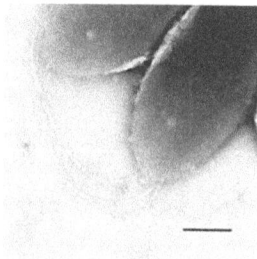

Polar flagella isolated from *Ruegeria* cells show polymorphic transition under acidic conditions. Filaments are either Coiled or Straight at pH 7.0; Straight between pH 4.0–6.2; a mixture of Straight and Normal; and filaments depolymerized below pH 3.0. Calculation of the helical diameter revealed that *Ruegeria* flagella belong to Family II, typical for the polar flagellum, despite its irregular morphogenesis.

Ruegeria sp. TM1040 genome (4,153,699 bp/ 3,864 genes)	NC_008044 CP000377 AAFG02000000

Multiple flagellin genes

motB flgEKLI-fliPNOFL---motA
(flaA) flhA fliR flhB --flgHAGF₁
fliQE flgCB fliI----flgF₂ (flbT)(flaF)
fliC3-flgJ fliK flgD

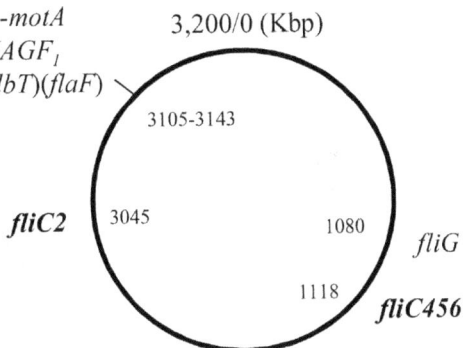

3,200/0 (Kbp)

3105-3143

fliC2   3045

1080   *fliG*

1118   *fliC456*

Missing genes: *flgM, fliA,D,H,M,S*

The genome of *Ruegeria* sp. TM1040 consists of one 3,200,938bp chromosome and two circular plasmids (0.8-Mbp and 0.1-Mbp). The majority of the flagellar genes form a compact cluster at 3,105 Kbp on the chromosome. The gene arrangement partly resembles that of the polar flagellum in *R. spaeroides* of the same alphaproteobacteria. There are **five flagellin genes (*fliC2-6*) on the chromosome** and one gene (*fliC1*) **on the megaplasmid** (not shown). No other flagellar genes are found on plasmids. Amino acid sequences of 6 flagellins considerably resemble one another. Only the *fliC3* gene is genetically necessary for swimming motility, suggesting that FliC3 is the major flagellin in polar flagellum.

Polar flagella isolated from *Ruegeria* cells give rise to two bands by SDS-PAGE: major 29 kDa and minor 27 kDa bands. From the amino acid sequencing of the N-terminus, FliC3 (282 aa) is 29 kDa flagellin, and any one of the others FliC1 (282 aa), FliC2 (282 aa), FliC4 (282 aa), or FliC5 (281 aa) can be 27 kDa flagellins.

Strains were provided by Robert Belas of the University of Maryland Biotechnology Institute, Baltimore, USA.

CHAPTER 23

Saccharophagus degradans — The Seaweed Eater

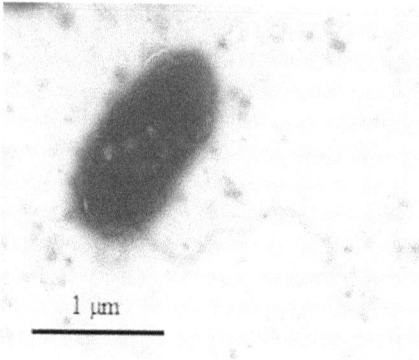

1 μm

Phylum	Proteobacteria
Class	Gammaproteobacteria
Order	Alteromonadales
Family	Alteromonadaceae
Genus	Saccharophagus
Species	*S. degradans*

Saccharophagus (formerly *Microbulbifer*) *degradans* is a Gram-negative, rod-shaped, aerobic, and motile bacterium. *S. degradans* is related to a group of marine bacteria that are able to **degrade a variety of complex polysaccharides** found in the ocean. The cell contains numerous vacuoles in the cytoplasm, which are called prosthecae, and releases them into the medium. The prosthecae has appendages that are neither pili nor flagella, but are extensions of the cellular membrane and contain cytosol.

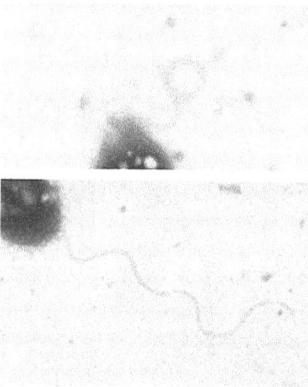

Each cell possesses a single polar flagellum, which is coiled at rest (left: upper) and extended at work (lower) as with the Fla1 flagellum in *R. sphaeroides* (see Chapter 21).

Agar is a complex polysaccharide extracted from marine red algae and used widely for microbiological culture media with solidity. *S. degradans* cells hydrolyze agar and develop depression of the surface of the agar plate. Prolonged incubation results in complete liquefaction of the agar surrounding the colonies.

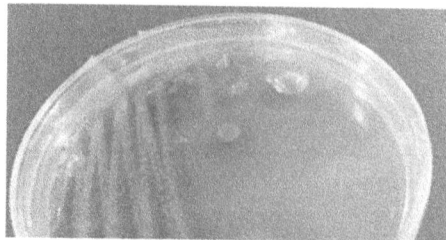

Saccharophagus degradans 2-40 genome (5,057,531 bp/ 4,007 genes)	NC_007912

Multiple flagellin genes

5,057/0 (Kbp)

Missing genes: *none*   *motBA*   4095

fliJIHGFE (motDC) mcp cheAZY fliS₁
flhGFA-B fliRQPONMLK-B cheRW-
fliJIHGFE --- (fleQ) fliS₂S₃D (flaG) fliC1-
fliC2 flgLKJIHGFEDCB cheRW flgAMN

2744-2800

2730

fliP' - - fliM'

The genome of the *Saccharophagus degradans* 2–40 strain consists of a single 5.0-Mbp chromosome. The flagellar genes form a compact cluster in a locus at the 2,744–2,800 Kbp region on the chromosome. This is **one of the most compact gene clusters**.

There is redundancy in several genes. There are two flagellins: FliC1 (587 aa) and FliC2 (580 aa). The identity between the two is 61.0%. There are two *fliM* genes: FliM' (119 aa) and FliM (333 aa). There are two *fliP* genes: FliP' (101 aa) and FliP (280 aa). There are three *fliS* genes: FliS₁ (250 aa), FliS₂ (130 aa), and FliS₃ (134 aa).

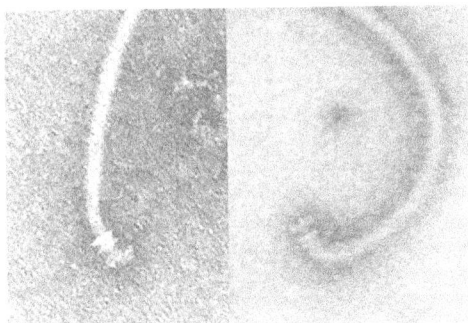

The flagellar basal body looks similar to that of *Salmonella*. The curvature of the hook is acute, which is different from the straight hooks of the polar flagellum in *G. oxydans* (Chapter 12) and *X. oryzae* (Chapter 29).

Strain was provided by Shogo Nakamura of Toyama University, Toyama, Japan. The project was carried out by Yuki Kato and Kohei Dono.

CHAPTER 24

Salmonella enterica Serovar Typhimurium — The Best-Studied Flagella

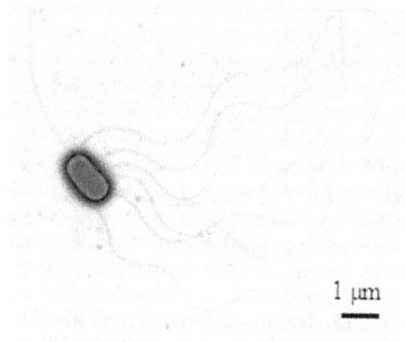

1 μm

Phylum	Proteobacteria
Class	Gammaproteobacteria
Order	Enterobacteriales
Family	Enterobacteriaceae
Genus	Salmonella
Species	*S. enterica*
Subspecies	Serovar Typhimurium

Salmonella enterica serovar Typhimurium (*S.* Typhimurium) is a Gram-negative rod-shaped bacterium. *S. enterica* serovar Typhi is the most frequent foodborne pathogen, while serovar Typhimurium is not harmful to humans but is to the mouse, as the name indicates, and thus is widely used as a lab strain. The flagellum of *S. typhimurium* is the best-studied, as well as or more than *Escherichi coli*, because of the abundance of their flagellar mutants (more than 100,000). Most of the structural studies have been carried out using the SJW strains (see Topic: History of *Salmonella* SJW strains).

The flagellar filament of *S.* Typhimurium is composed of thousands of subunits of a single kind of protein called flagellin. Because of the simplicity of the system, several important discoveries were made in the early days of research.[1-4]

1. A filament has a **polarity in the structure**. The proximal end of a filament appears as an arrowhead, while the distal end looks like a fish tail as observed by electron microscopy. Later on, the explanation for this came from the atomic structure of flagellin. Flagellins of elongated shape pack cylindrically at a tilted angle on the crystal lattice of the filament.
2. A filament grows distally (**distal growth**); nascent flagellin subunits are added to the tip of the existing filament.
3. Filaments are **heat- and acid-sensitive**; they depolymerize into monomeric flagellin by heat treatment at 60°C or by acid treatment at pH 2.5.
4. Monomers polymerize into filaments in the presence of short-seed fragments of filaments under physiological conditions **in vitro**, demonstrating that flagellar filament is **a self-assembly system**.
5. **In vivo**, a filament elongates with the aid of **the cap protein** (FliD or HAP2). Without the cap protein, monomeric flagellin is secreted into the medium.
6. Flagellins **copolymerize** with other flagellins among the same family, but not with those that belong to other families.[5]

Salmonella enterica **serovar Typhimurium LT2 genome** **(4,951,371bp/ 4,558genes)**	NC_003197 AE006468

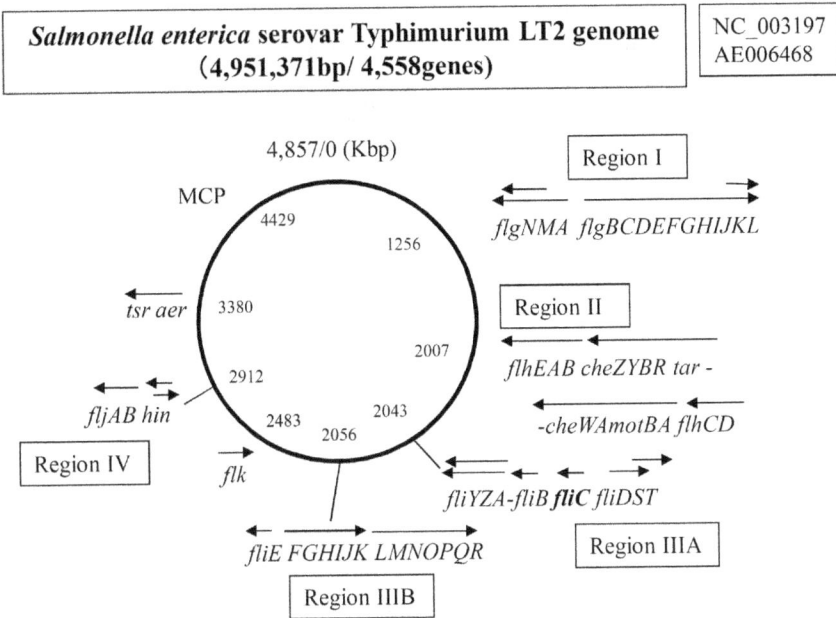

The genome of *S.* Typhimurium LT2 consists of one 4.8-Mbp chromosome and one 93-Kbp plasmid. Flagellar genes form compact clusters in five regions on the chromosome. Each region contains flagellar genes from the different classes of the flagellar transcriptional hierarchy (see Topic: Gene regulation).

Most of the flagellar proteins are secreted through the so-called type III secretion system (T3SS). Only three secreted proteins (FlgA, FlgH, and FlgI) possess type II signal peptides and are secreted through the Sec general secretion pathway into the periplasmic space to form the PL ring complex. Four operons (*fliAZY*, *flgMN*, *fliDST*, and *flgKL*) are expressed from both class 2 and class 3 promoters of the transcriptional hierarchy (double arrows).

Although the presence of an N-terminal secretion signal is required for secretion, there are no consensus sequences found in the N-terminal region of flagellar proteins secreted from T3SS. Many of them have the N-terminus *f*-Met cleaved off before or during secretion, for unknown reasons.

There are several local (species-specific) genes that are found only in *Salmonella* and related species: *flhCD*, *fliYZ*, *fliB*, *fliO*, *flk*, *fljAB*, and *hin* (see Introduction).

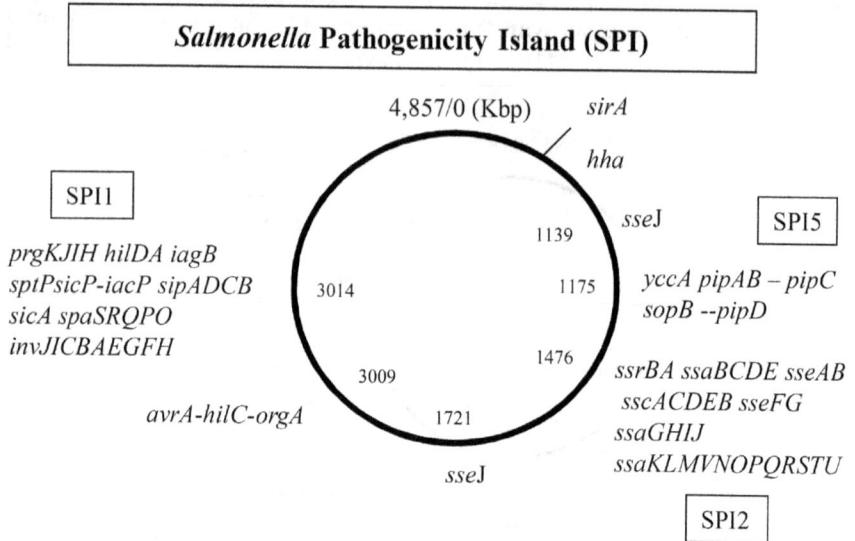

Salmonella Pathogenicity Island (SPI)

4,857/0 (Kbp)   *sirA*

hha

SPI1

*prgKJIH hilDA iagB
sptPsicP-iacP sipADCB
sicA spaSRQPO
invJICBAEGFH*

3014

3009

avrA-hilC-orgA

1721

sseJ

1139

1175

1476

sseJ

sseJ

SPI5

*yccA pipAB – pipC
sopB --pipD*

*ssrBA ssaBCDE sseAB
sscACDEB sseFG
ssaGHIJ
ssaKLMVNOPQRSTU*

SPI2

In the *Salmonella* genome, there are at least three gene clusters that are related with pathogenicity. They are called Salmonella Pathogenicity Islands and numbered: SPI1, SPI2, and SPI5. SPI1 is necessary for invasion upon the first contact with the host. SPI2 is required for surviving in macrophages after invasion. The virulence secretion apparatus of SPI1 is called the needle complex. SPI5 includes effectors which are secreted from the SPI1 needles.

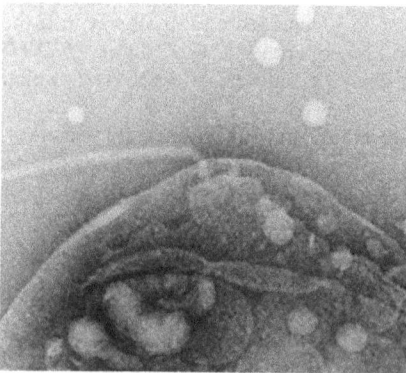

The SPI1 genes that are under control of the *hil* regulatory genes are induced in the presence of salts, e.g., 100 mM NaCl.[6] Judging from gene homology and the resemblance of structures, the needle structure seems to share a common ancestor with the flagellar structure (see Topic: Flagella and pathogenicity).

The SPI2 genes have high homologies with those of the SPI1 needle complex, but its structure has not been identified.

Strains were provided by Shigeru Yamaguchi, Saitama, Japan, and Kelly Hughes of the University of Utah, Salt Lake City, USA.

History of *Salmonella* SJW Strains

In the early days of flagella research, flagellar mutants of SJW strains were most frequently used to establish basic knowledge of the flagellum. Here is the historical background to show where SJW strains were derived from.

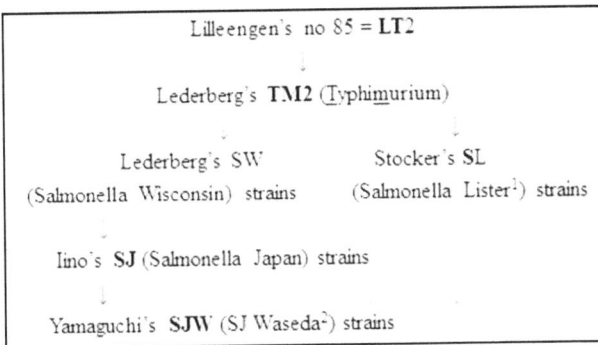

```
              Lilleengen's no 85 = LT2
                        ↓
           Lederberg's TM2 (Typhimurium)
                ↓                    ↓
      Lederberg's SW          Stocker's SL
  (Salmonella Wisconsin) strains   (Salmonella Lister¹) strains
                ↓
      Iino's SJ (Salmonella Japan) strains
                ↓
   Yamaguchi's SJW (SJ Waseda²) strains
```

1. Stocker worked at the Lister Institute of Preventive Medicine, informally known as the Lister Institute, in Chelsea in London before he moved to Berkeley.
2. Yamaguchi worked as a part-time lecturer of science education at Waseda University and collected his strains in the corner of the student lab throughout his lifetime.

Tetsuo Iino learned flagellar genetics from Lederberg, introduced it in Japan in 1962, and named his strains SJ (Salmonella Japan). Shigeru Yamaguchi, an ex-student of Iino, has collected flagellar mutants (more than 100,000!) throughout his lifetime and named the stable mutants as SJW (Salmonella Japan Waseda) strains. The total number of SJW strains is 3,334 as of 2013, which is the largest single collection of flagellar spontaneous mutants.

SJW strains are all spontaneous mutants, and the genetic map of flagellar genes was made by P22-mediated transduction in Yamaguchi's time. Consequently, the genetic map is not on a physical scale but on a functional scale. For example, the flagellin (*fliC*) gene is divided into 22 sections, each of which includes 5–10 Fla⁻ mutants. Many of them have not been used and await further analysis.

In the genome era, Yamaguchi's method is old fashioned. Today deletion mutants of *E. coli*/*Salmonella* are systematically constructed and analyzed using sophisticated machines. It is like a fisherman with a rod fighting against a fish vs. a big fishing boat with a big fishing net scooping everything in the sea. This may just be nostalgia, but I love the way of Yamaguchi.

CHAPTER 25

Selenomonas ruminantium — The Authentic Lateral Flagella

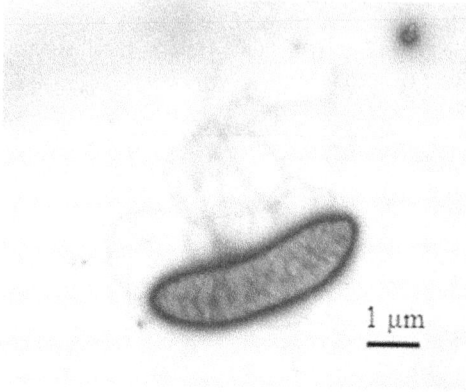

1 μm

Phylum	Firmicutes
Class	Negativicutes
Order	Selenomonadales
Family	Veillonellaceae
Genus	Selenomonas
Species	*S. ruminantium*

Selenomonas ruminantium is an **anaerobic bacterium**, often found in the rumen of domestic animals. *S. ruminantium* is one of a few **Gram-negative bacteria among the phylum Firmicutes**, most of which are Gram-positive bacteria. The cells are kidney- or crescent-shaped. The flagellar formation is suppressed in the presence of glucose, indicating that the flagellar system is regulated under **catabolite repression**. *S. ruminantium* is the sole species so far found that produce **authentic lateral flagella**, as Leifson mentioned in his "Atlas of Bacterial Flagellation" (see Topic: Flagellar position and shape).

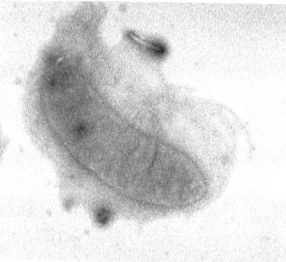

S. ruminantium cells produce lateral flagella (a tuft of flagella near the midpoint of the cell body), but nevertheless they swim along the cell's long axis. Other species that are reported to have lateral flagella include *V. parahaemolyticus*, *V. alginolyticus*, *B. japonicum*, *Photobacterium profundum*, and *R. sphaeroides*. All of these species produce either polar or lateral flagella or both, depending on growth conditions. In contrast, *S. ruminantium* cells possess several lateral flagella, but no polar flagella. The growing position of flagella is localized near the midpoint of the cell body, indicating they are authentic lateral flagella, whereas others are actually peritrichous (see Introduction and Topic: Flagellar position and shape).

The flagellum is composed of a single kind of flagellin with the molecular weight of 71 kDa by SDS-PAGE. The flagellin is heavily glycosylated, and the N-terminus is blocked.

Selenomonas ruminantium subsp. lactilytica TAM6421 genome (3,631,933 bp/ 3,512 genes)

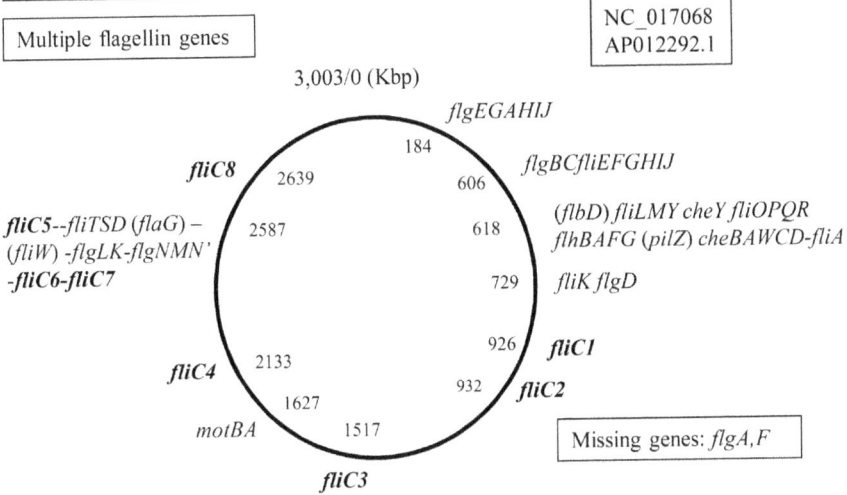

Multiple flagellin genes

NC_017068
AP012292.1

3,003/0 (Kbp)

flgEGAHIJ

fliC8

184

flgBCfliEFGHIJ

2639

606

*fliC5--fliTSD (flaG) –
(fliW) -flgLK-flgNMN'
-fliC6-fliC7*

2587

618

*(flbD) fliLMY cheY flioPQR
flhBAFG (pilZ) cheBAWCD-fliA*

729

fliK flgD

926

fliC1

fliC4

2133

932

fliC2

1627

motBA

1517

Missing genes: *flgA,F*

fliC3

The genome of *S. ruminatium* TAM6421 strain consists of one 3.0-Mbp chromosome and nine plasmids. The total number of nucleotides is 3,631,933 bp. Flagellar genes are clustered in several sites on the chromosome. The gene arrangement is the same as that of *Paenibacillus* sp. There are **eight *fliC* genes** but only one major band was detected at 71 kDa from purified filaments by SDS-PAGE, which corresponds to *fliC3*. The function of the other flagellins is unknown.

Urgent proposal for an interesting project

The *flhFG* genes—usually required to place flagella at a pole—may be used for placing flagella at the cell side. Remove the *flhFG* genes and see if the flagella become peritrichous or not. If not, there might be a chance to discover a new gene(s) for flagella placement.

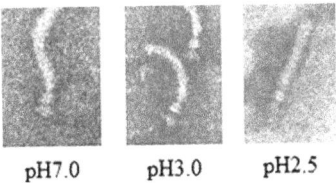

pH7.0   pH3.0   pH2.5

The hook length of *S. ruminantium* (105 ± 12 nm) is almost double the *Salmonella* hook length (55 ± 6 nm). This is **the longest hook** in nature so far studied (see Topic: Hook length). The shape of the hook is S-shape at pH 7.0, arc at pH 3.0, and straight at pH 2.5.

Strains were provided by Naoki Abe of Tohoku University, Sendai, Japan.

Hook length

The phenomenon indicating that the hook length is controlled by a gene was discovered about 40 years ago. Silverman and Simon (1972) found that a *flaE* (the previous name of *fliK* in *E. coli*) mutant produced extraordinarily long hooks, which they named polyhooks,[1] and concluded that the *flaE* gene "functions to control the length of the hook." Around the same time, Patterson-Delafield et al. (1973) found that a *flaR* (the previous name of *fliK* in *S. typhimurium*) mutant similarly produced extraordinarily long hooks, which they named superhooks[2] (see Introduction: Unified gene names).

The hook length is controlled by *fliK*.[3] Additionally, the needle length is controlled by *fliK* homologs.[4,5] As seen in the table below, the hook/needle length is proportionate to the size of N-terminal domains of FliK, indicating that the molecular mechanism might be the same for both systems.[6]

Table T8.1 Sizes of FliK and the Hook Length					
FliK & homologs	**Species**	**Molecular Size (a.a.)**	**Size of N-term domain (a.a.)***	**Hook or needle Av. Length (nm)**	**Ref.**
FliK	*S. ruminantium*	817	670	105 ± 12	7
FliK	*R. sphaeroides*	699	575	71 ± 6.7	8
Yscp	*Y. enterocolitica*	515	403	58	9
YscP	*Y. pestis*	455	342	41	4
FliK	*S. typhimurium*	405	269	55 ± 6.0	3
InvJ	*S. typhimurium*	336	267	50	10
Spa32	*S. flexneri*	292	223	45	11

The sizes of N-terminal domains of FliK and the homologs are estimated by deducing the sizes of the C-terminal conserved domain from the molecular sizes.

The regulation of hook length is never tight, but is rather loose. Assembly of hook/needle lengths apparently shows Gaussian distribution. The length distribution of *Salmonella* hooks is: 55 ± 6.0 nm. The hook length measured includes the HAP region. Actual length of the hook portion is 45 nm.[12]

The molecular mechanism of the hook length control is still controversial.[13–16] In the most popular model, the physical ruler directly measures hook length during its secretion through the central hole of the hook. It is unclear how the N-terminus transmits a signal to the C-terminal domain to switch the substrate specificity. The atomic structure of FliK shows that the C-terminal domain has a compact stable structure, which interacts with FlhB, the secretion gatekeeper. In contrast, the N-terminal region does not have stable structures, but is fluctuating.[17] Since FliK is involved in gene regulation (seen Topic: Gene regulation), it is most important to elucidate the structure–function relationship of the N-terminal region.

Multiple Flagellins

As the number of genomes solved increased, it turned out that numerous species have multiple flagellin genes in their mapped genomes. Here are some examples: there are six (*flaA–flaF*) in *A. fisheri*, six (*fliC1–fliC6*) in *B. bacteriovorus*, seven (*fljJ–fljO, fliC*) in *C. crescentus*, six (*fliC1–fliC6*) in *Ruegeria* TM1040, and eight (*fliC1–fliC8*) in *S. ruminantium*. The position of each flagellin is known in *B. bacteriovorus* and *C. crescentus*.

B. bacteriovorus possesses a single polar flagellum that appears in a tapered wave. Flagellins are located in separate positions in a filament; a small amount of FliC3 at the proximal end, followed by a large amount of FliC5, a large amount of FliC1, a small amount of FliC2 in this order, and a large amount of FliC6 at the distal end. FliC4 is present at a low level, but the location has not been determined.

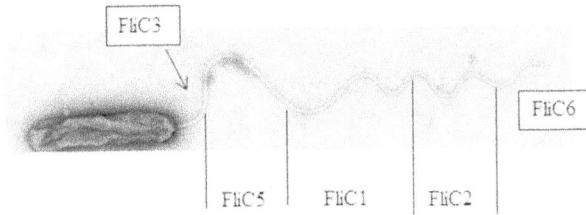

C. Crescentus possesses a single polar flagellum. Driks et al. (1989) did pioneering work; they determined or inferred the positions of six flagellins, using antibodies against several of them. Electron microscopic images of immunostained filaments revealed that each flagellin locates in separate positions of the filament. FliK, M, N, and O are homologous in sequence/size (273 aa) and are immunologically indistinguishable.

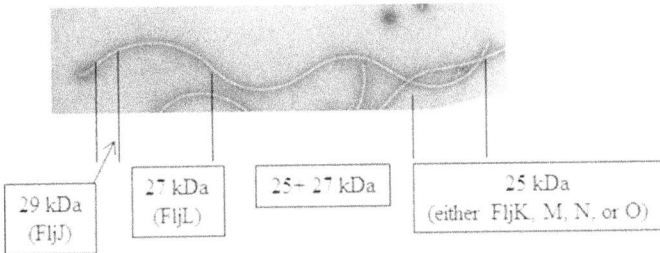

However, in general, the roles of the flagellin isomers have not been clarified, because deletion of any of the isomer genes usually does not affect the filament shape or motility.

CHAPTER 26

Sinorhizobium meliloti — Nitrogen–Fixer in the Grassland

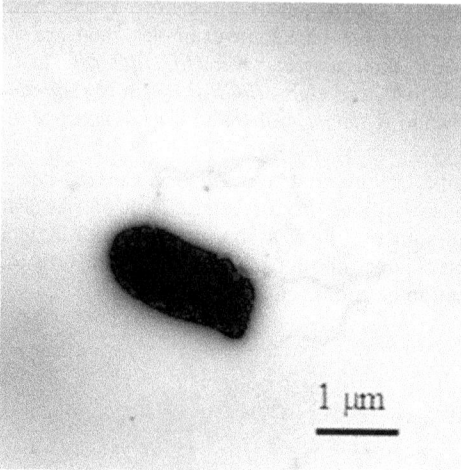

1 μm

Phylum	Proteobacteria
Class	Alphaproteobacteria
Order	Rhizobiales
Family	Rhizobiaceae
Genus	Sinorhizobium
Species	*S. meliloti*

Sinorhizobium (formerly *Rhizobium*) *meliloti* is a Gram-negative rhizosphere bacterium, best known for its ability to induce the formation of nodules on the roots of legumes such as alfalfa (*Medicago*) and sweet clover (*Melilotus*). **Bacteroids in the nodules fix atmospheric nitrogen** and leave excess nitrogen behind to the benefit of the plant.

Each cell possesses a few peritrichous flagella and **swims in liquid only when the cell density is low.**[1] This may be related to the fact that *S. meliloti* cells must transit from the aerobic cycle in the soil to the microaerobic cycle in the nodule environment.

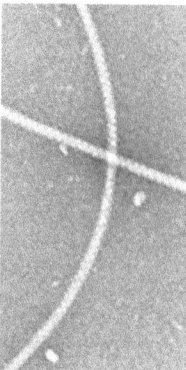

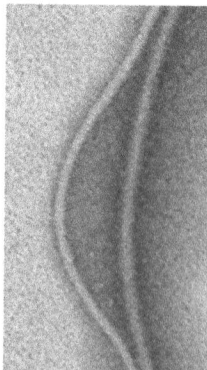

The *S. meliloti* filament (left) is called a **complex filament** in comparison with the plain filaments of *B. japonicum* (right), because it characteristically shows a zigzag pattern on the surface.[2,3]

Sinorhizobium meliloti SM11 genome (7,173,736 bp/ 7,093 genes)	NC_017325 CP001830

Multiple flagellin genes

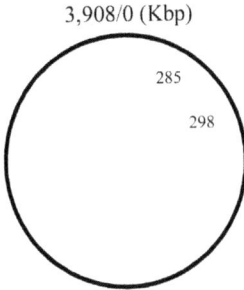

3,908/0 (Kbp)

285

298

mcpE – cheY₁AW₁RB₁Y₂D –fliF- visNR

mcpE – cheY|AW|RB|Y₂D –fliF- visNR

flhB fliGNM motA – flgF flil – flgBC fliE flgGAI (motE) flgH fliLP fliC1C2 – fliC3C4 – motBC flik – rem - flgEKL flaF (flbT) flgD fliQ flhA fliR

Missing genes: *flgJ,M,N, fliA,D,H,J,S*

The genome of the *Sinorhizobium meliloti* SM11 strain consists of one 3,908,022 bp chromosome and two megaplasmids (1,633,319 bp and 1,632,395 bp). The majority of the flagellar genes form a compact cluster at 298 Kbp on the chromosome. Although this is one of the most densely packed clusters of the flagellar genes (see Topic: Gene arrangement), the gene arrangement is as random as that of the thin flagellum in *B. japonicum* (see Chapter 7). The *S. meliloti* genome contains **four flagellin** genes, *fliC1C2–fliC3C4*. Genes required for nodule formation (*nod*) are localized on both plasmids. MotE is chaperone-specific for the periplasmic motility protein MotC.

VisN and **VisR** form a heterodimer acting as the global regulator ("master switch") of *fla*, *che*, and *mot* genes in *S. meliloti*.[4] These proteins are the counterpart of FlhCD in *S.* Typhimurium.

Rem (for *"regulator of exponential growth motility"*) is a unique autoregulatory protein responsible for exclusive motility during exponential growth. This function is the molecular explanation for the restriction of motility to low cell density.[1]

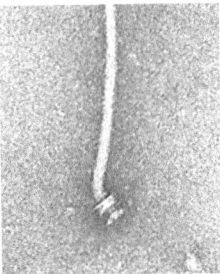

There are **four flagellin** genes, all of which produce functional flagellin monomers, although the Fla3 (321 aa) is shorter than the other three (395 aa). FliC1 and FliC2 are 92% homologous, which is **the closest profiles** among multiple flagellins. Complex flagella are composed of flagellin dimers. Deletion analysis has revealed that **FliC1** is the principal, absolutely essential subunit, but that in addition all or at least one of the secondary flagellin species: **FliC2, FliC3, or FliC4**, are required for the assemblage of an intact filament.[5]

Strains were provided by Rudiger Schmitt of University of Regensburg, Regensburg, Germany.

CHAPTER 27

Symbiobacterium thermophilum — A Gram-Negative, High (G+C) Firmicutes

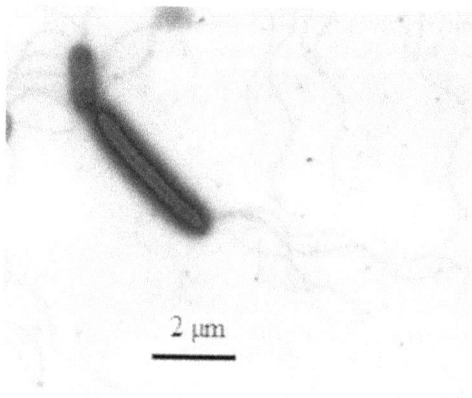

2 μm

Phylum	Firmicutes
Class	Clostridia
Order	Clostridiales
Family	Clostridiales Family XVIII. Incertae Sedis
Genus	Symbiobacterium
Species	S. thermophilum

Symbiobacterium thermophilum is one of a few **Gram-negative and high (G+C) bacteria in the phylum Firmicutes,** most of which are Gram-positive and low (G+C) bacteria (see also *Selenomonas*). A mixed culture of *S. thermophilum* and *G. stearothermophilus* was originally obtained from a compost as a result of its thermostable tryptophanase activity. Naturally, *S. thermophilum* is characterized by its syntrophic mode of growth, effectively growing in a coculture with *Geobacillus stearothermophilus* strain S, but not in its pure culture. The ecological survey has revealed that the group of bacterium widely occurs in the natural environment including compost, soil, animal intestine, and sea water. *G. stearothermophilus* supports the growth of *S. thermophilum* due to its multiple functions supplying growth promoting substances including CO_2 and amino acid(s) and eliminating growth inhibitory activities. The requirement for CO_2 may be due to the lack of carbonic anhydrase, which catalyzes interconversion between CO_2 and bicarbonate. The optimum growth temperature is 60°C.

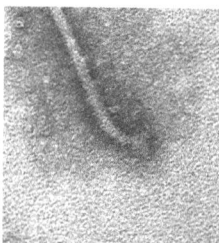

The flagellar basal body of *S. thermophilum* shows only two rings, a typical structure for Gram-positives, though Gram-stain gives a negative reaction due to the distinctive multi-layered cell wall.

Symbiobacterium thermophilum IAM 14863 genome (3,566,135 bp/ 3,337 genes)	NC_ 006177

3,337/0 (Kbp)

motBA fliC – fliSD (flaG) flgLK – flgM –
fliA flhGFAB fliRQPON'L flgED fliK –
fliJIHGFE flgCBG cheX₂ fliM

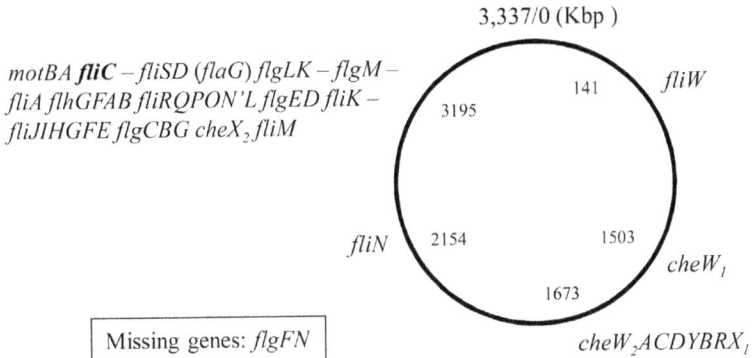

fliW

141

3195

fliN   2154   1503

1673   *cheW₁*

Missing genes: *flgFN*

cheW₂ACDYBRX₁

The genome of the *Symbiobacterium thermophilum* IAM 14863 strain consists of a single 3.3-Mbp chromosome. The majority of the flagellar genes form a compact cluster at 3,195 Kbp on the chromosome. The *fliO* gene is not essential, but is found in *E. coli*/*Salmonella*, *A. missouriensis*, *A. fischeri*, *P. aeruginosa*, and *R. solanacearum*.

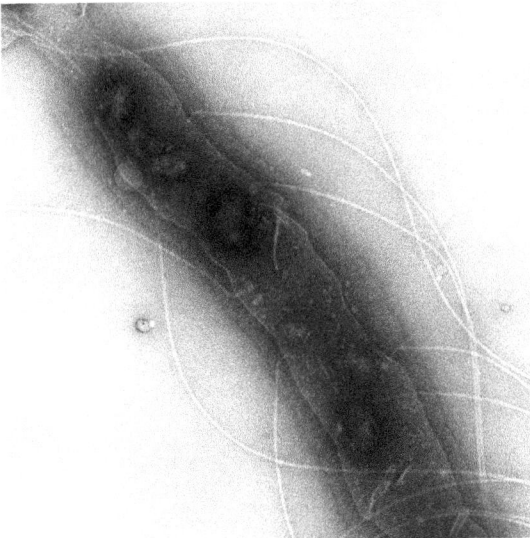

Each cell grows either subpolar flagella (previous page) or peritrichous flagella (left), depending on the growth phase. Since there is only one flagellar system in this organism, flagella positions must be switched on-off in each cell. Whether the *flhFG* genes are involved in this event awaits direct proof.

Strain was provided by Kenji Ueda of Nihon University, Fujisawa, Japan.
The project was carried out by Kaoru Uchida.

CHAPTER 28

Vibrio parahaemolyticus — Polar/Lateral Flagella with H+/Na+ Motor

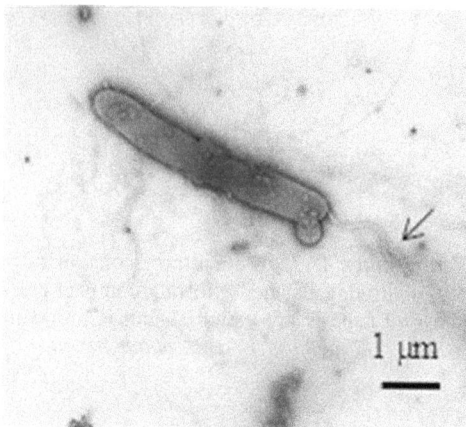

Phylum	Proteobacteria
Class	Gammaproteobacteria
Order	Vibrionales
Family	Vibrionaceae
Genus	Vibrio
Species	*V. parahaemolyticus*

Vibrio parahaemolyticus is a Gram-negative marine bacterium. The organism is a pathogen of humans and often causes food-borne illness such as gastroenteritis. The cell possesses two types of flagella:[1] **numerous lateral flagella and a single polar flagellum** (arrow). Lateral flagella are induced under viscous environments and promote swarming motility on surfaces,[2] while the polar flagellum is expressed continuously and propels swimming motility.[3,4] The polar flagellum is powered by **the sodium motive force**, whereas the lateral flagella are powered by **the proton motive force**.[5] Having all possible flagellar systems, *V. parahaemolyticus* is regarded as a model organism for dual flagella systems. The polar flagellum rotates much faster than proton-powered motors, and the sodium-driven motor of the related *Vibrio alginolyticus* has been elegantly deciphered to reveal an unusual basal body structure containing two additional ring structures.[6,7]

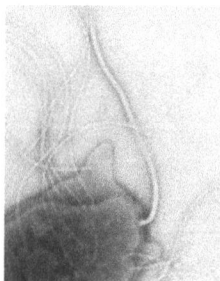

The polar flagellum appears thicker than the lateral flagellum due to the sheath, which is physically fragile (left). The thickness of the polar filament itself looks similar to that of the lateral filament. However, the helix pitch of the polar filament is 50% greater than that of the lateral filament. They belong to different flagellar families (see Appendix: Flagellar family). There are six highly homologous flagellins for the polar flagellum. Mutational and protein analyses suggest all of these can be produced and incorporated into a filament and none are essential. The single lateral flagellin (284 aa) is smaller than the polar flagellins (374–384 aa).

Vibrio parahaemolyticus **BB22OP genome** (5,103,524 bp/ 4,447 genes)

Multiple flagellin genes

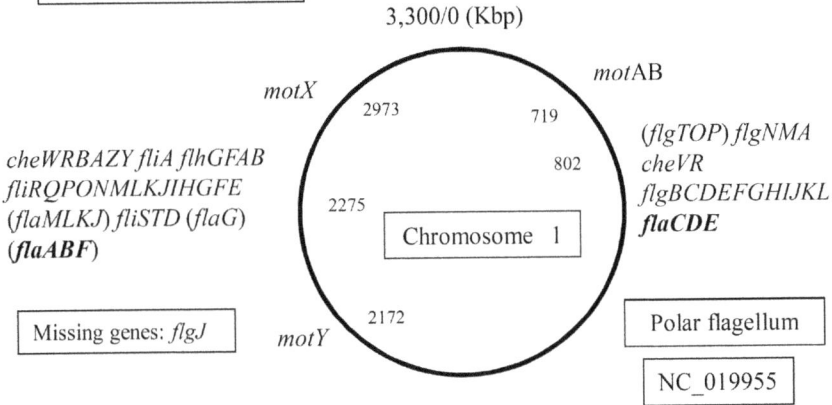

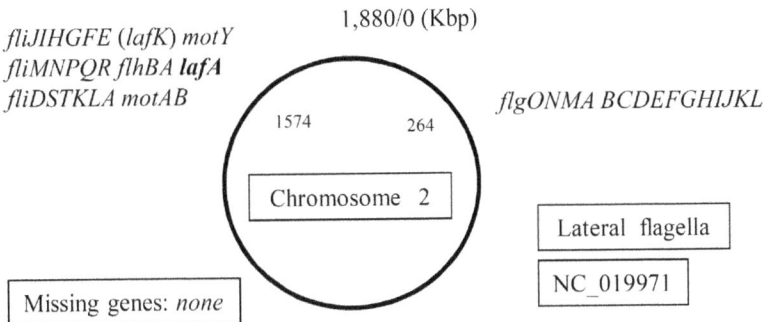

3,300/0 (Kbp)

motX

*mot*AB

cheWRBAZY fliA flhGFAB
fliRQPONMLKJIHGFE
(flaMLKJ) fliSTD (flaG)
(flaABF)

(*flgTOP*) *flgNMA*
cheVR
flgBCDEFGHIJKL
flaCDE

2973
719
802
2275
Chromosome 1

Missing genes: *flgJ*

2172

motY

Polar flagellum

NC_019955

---

fliJIHGFE (lafK) motY
fliMNPQR flhBA lafA
fliDSTKLA motAB

1,880/0 (Kbp)

flgONMA BCDEFGHIJKL

1574
264
Chromosome 2

Lateral flagella

NC_019971

Missing genes: *none*

The genome of the *Vibrio parahaemolyticus* BB22OP strain consists of two chromosomes: chromosome 1 (3,297,305 bp) and chromosome 2 (1,806,219 bp). Flagellar genes for the polar flagellum form compact clusters on chromosome 1, while those for the lateral flagella form them on chromosome 2. The gene arrangement is similar to that of *Salmonella/E. coli*. There are **six polar flagellins** (*fla*A–F) on chromosome 1 and **one lateral flagellin** (*laf*A) on chromosome 2. Polar flagellins are homologous in size (374–384 aa) and in a.a. sequence (66–77%). Lateral flagellins are homologous to polar flagellins, but shorter (284 aa). There are three new genes: *flgTOP.flgT* encodes protein for T ring,[8] *flgO* is murein transglycosylase, and *flgU* has an unknown function but is required for flagellar assembly.

Strains were provided by Linda McCarter of the University of Iowa, Iowa City, USA.

CHAPTER 29

Xanthomonas oryzae pv. *Oryzae* — Pathogen in the Rice Country

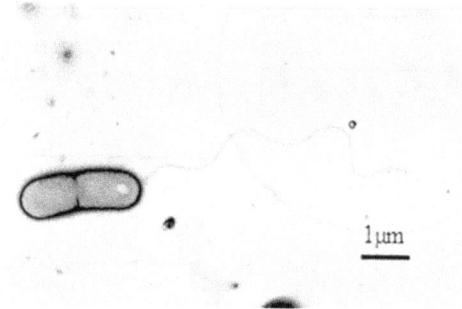

Phylum	Proteobacteria
Class	Gammaproteobacteria
Order	Xanthomonadales
Family	Xanthomonadaceae
Genus	Xanthomonas
Species	*X. oryzae*

1 μm

Xanthomonas oryzae is a Gram-negative, aerobic, plant pathogen the is the main cause of blight and leaf streak of rice (*Oryza* spp.). *X. oryzae* cells produce an extracellular acidic heteropolysaccharide called xanthan, and thus its colonies are yellow. Cells are relatively electron transparent, and cytoplasmic structural components are visible. Each cell possesses a single polar flagellum. Flagellation is suppressed in the presence of glucose, indicating that flagellar genes are regulated by catabolite repression.

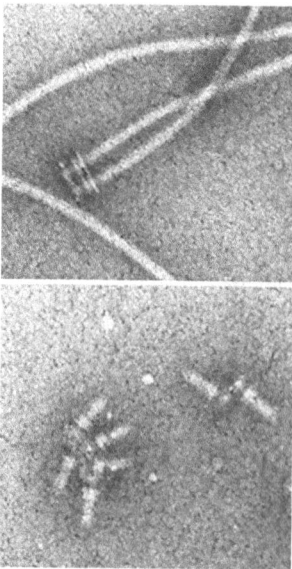

The flagellar basal body of *Xanthomonas oryzae* resembles that of *Salmonella*. However, the hook is straight (left) that is often observed with polar flagella or with short hooks. When the hook basal body is isolated (below), the average hook length is 35 nm, much **shorter** than the *Salmonella* hook (55 nm). This is a mystery because the hook-length controller FliK (435 aa) is longer than *Salmonella* FliK (405 aa).

Xanthomonas oryzae pv. **Oryzae KACC 10331 genome**
(4,941,439 bp/ 4,065 genes)

NC_006834

4,941/0 (Kbp)

72

709

T3SS
Virulence genes

motBA

2774

flhBAFG fliA cheYZA

2753

fliEFGHIJKLMNOPQR

2738

(rpoN) *(fleQ)*

2718

flgMA (cheV) flgBCDEFGHIJKL fliCDS

Missing genes: *flgN*

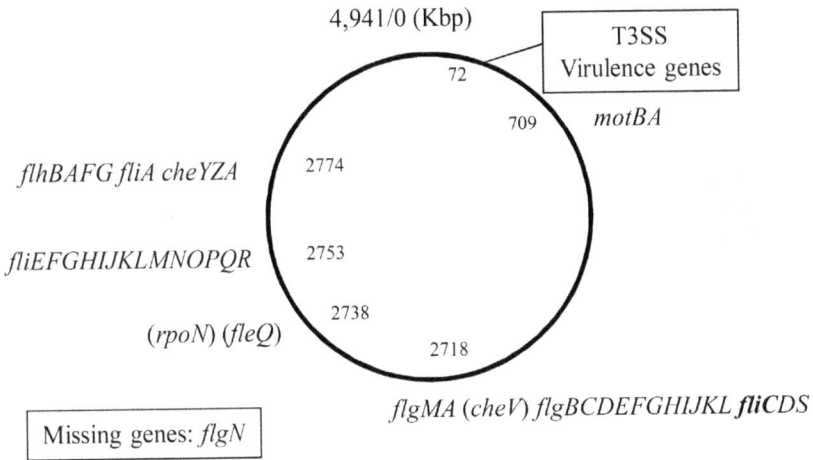

The genome of the *Xanthomonas oryzae* pv. oryzae KACC 10331 strain consists of a single 4.9-Mbp chromosome. Flagellar genes form compact clusters in three loci on the chromosome in a similar manner to those of *Salmonella*. *X. oryzae* has a type III virulence secretion system, whose genes (*hrc*, *hrp*, *hpa*) form a compact cluster in a locus at 72 Kbp. The secretion apparatus has not been identified.

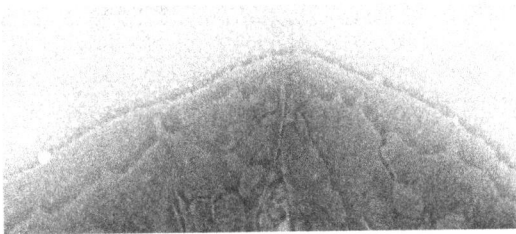

On the cell surface of *Xanthomonas oryzae* cells, there are numerous structures that connect the inner and outer membranes, suggesting the existence of a secretion apparatus.

Strains were provided by Fang-Sik Che of Nagahama Institute of Bio-Science and Technology, Nagahama, Japan.
The project was carried out by Yoshika Nosaka and Norio Yamamoto.

CHAPTER 30

Uncharacterized species — Slowly-Growing Bacteria

The number of bacterial species studied in the lab are limited due to the special culture conditions required for growth. There are many species that are known to exist in nature, but are not culturable in the lab, the so-called "viable but non-culturable (VBNC)" category. They will eventually be culturable once we learn what is missing for them to grow, as shown with *Symbiobacterium* (Chapter 27). There is another group of bacterial species that are culturable, but grow very slowly.

Here I show some species of the slowly-growing bacteria isolated by Prof. Tsutomu Hattori from paddy soil. These Hattori strains grow in a diluted culture medium (e.g., a 100× dilution of the nutrient broth) at room temperatures (10–25°C) without active aeration. It takes only a few days for some strains, but takes a week or longer for others to grow to a cell density high enough to observe by microscopy.

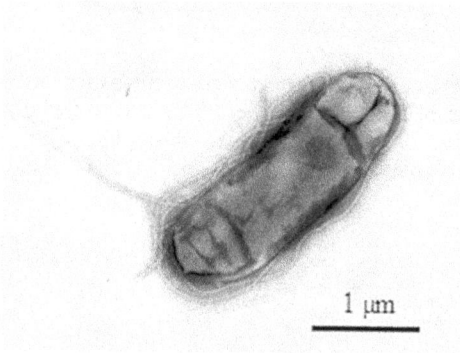

1 μm

Phylum	Proteobacteria
Class	Alphaproteobacteria
Order	Rhizobiales
Family	Bradyrhizobiaceae
Genus	Bradyrhizobium
Species	*B. oligotrophica*

Bradyrhizobium (formerly *Agromonas*) *oligotrophica* S58 was isolated from the paddy soil in the experimental station of Tohoku University in Sendai. *B. oligotrophica* is Gram-negative, aerobic, irregularly rod-shaped and salt-sensitive (unable to grow in the presence of 0.4% NaCl). The cell has polar flagella, which look soft and kinky judging from the appearance of flagella surrounding the cell body. Nevertheless, the cell can swim smoothly in liquid.

Bradyrhizobium oligotrophica S58 genome (8,264,165 bp/ 7,228 genes)	NC_020453 AP012603

Multiple flagellin genes

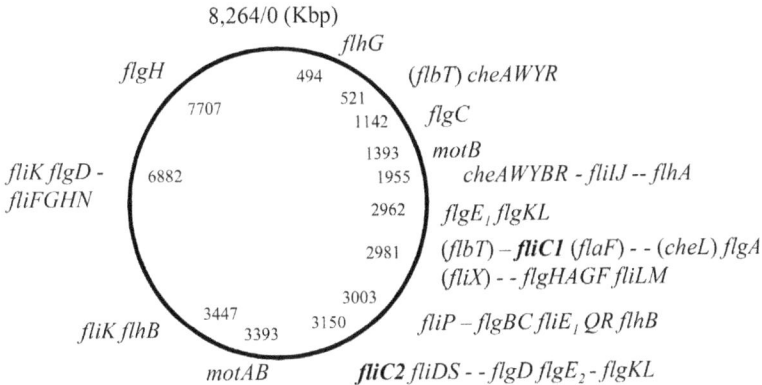

8,264/0 (Kbp)

flhG

flgH   494   (*flbT*) *cheAWYR*

7707   521   *flgC*
                   1142

fliK flgD -   1393   *motB*
fliFGHN   6882   1955   *cheAWYBR - fliIJ -- flhA*

2962   *flgE₁ flgKL*

2981   (*flbT*) – *fliC1* (*flaF*) - - (*cheL*) *flgA*
                          (*fliX*) - - *flgHAGF fliLM*

3003

3447   3150   *fliP – flgBC fliE₁ QR flhB*
fliK flhB   3393

motAB   *fliC2 fliDS - - flgD flgE₂ - flgKL*

Missing genes: *flgI,J,M,N, fliA,E,Q,R*

The genome of *Bradyrhizobium* (formerly *Agromonas*) *oligotrophica* S58 strain consists of a single 8.2-Mbp chromosome. The flagellar genes are scattered all over the chromosome and arranged as randomly as those in *H. pylori* (see Topic: Randomness of gene arrangement). There are several gene duplications. Two flagellins (523 aa and 274 aa) retain the conserved sequences in both terminal regions. So do the two hook proteins (600 aa and 412 aa).

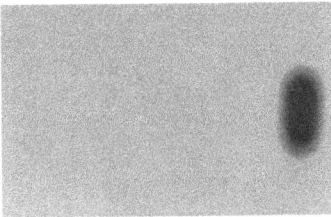

1 μm

Sphingomonas Japsi S23419
It takes more than 10 days to recognize its growth. The cell size is normal. It has a single straight flagellum at a subpolar position and is thus nonmotile.

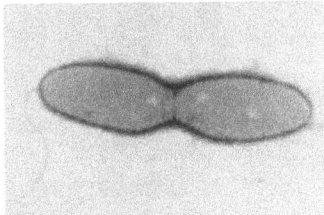

Zoogloea sp. S22235
It takes 2 days to grow. This species has a huge cell body and a polar flagellum at only one pole. The cell shows the switch-backing motility.

Strains were provided by Tsutomu Hattori of Tohoku University, Sendai, Japan. The project was carried out by Keisuke Matsuzaki.

CHAPTER 31

Buchnera aphidicola — Flagella Not for Motility

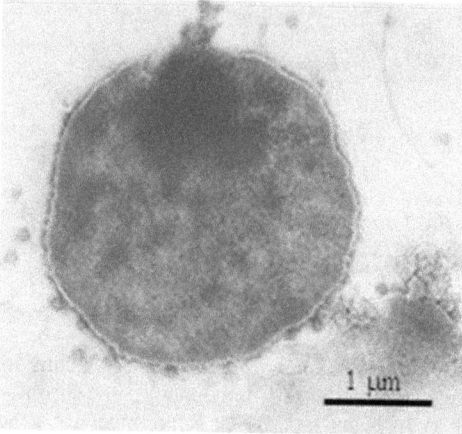

Phylum	Proteobacteria
Class	Gammaproteobacteria
Order	Enterobacteriales
Family	Enterobacteriaceae
Genus	Buchnera
Species	*B. aphidicola*

Buchnera aphidicola APS is the primary endosymbiotic bacterium of the pea aphid, *Acyrthosiphon pisum* (Harris), and the bacterial cells are harbored by large, differentiated cells called "**bacteriocytes**" in the fat body. The *Buchnera* genome is **about 1/7 of the *E. coli*** genome and is lacking many genes of metabolic enzymes necessary for free living. Because the cells do not grow on an artificial culture medium, they are collected from the bacteriocytes for microscopic observation. *Buchnera* cells harbor **hundreds of flagellar basal bodies** on the membrane, but do not produce filaments.

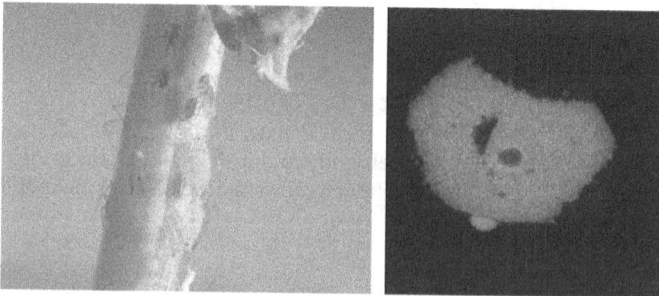

The pea aphid, *Acyrthosiphon pisum* (Harris), on the pea stem (left), and a bacteriocyte by confocal microscopy (right). DNA in a bacterioycyte was stained with TO-PRO-3. The cytoplasm is filled with *Buchnera* cells around the huge polyploid nucleus. Photos were provided by Shuji Shigenobu.

Buchnera sp. APS genome	NC_002528
(655,725 bp/ 574 genes)	BA000003

640/0 (Kbp)

fliE FGHIJK-MNPQR

76

369   268

flgNA BCDEFGHIJK   *flhBA*

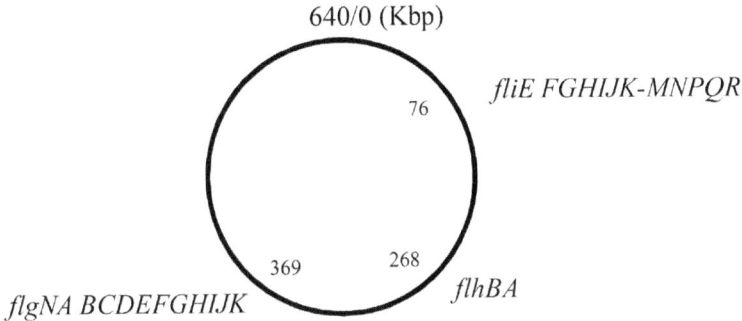

Missing genes: *fliA,C,D,L,S flgLM*

The genome of *Buchnera* sp. APS consists of one 640,681 bp chromosome and two plasmids (7,258 bp and 7,786 bp). Flagellar genes required for assembly of the basal structure up to the hook are clustered at two sites on the chromosome, and they are aligned in the same order as those in *Salmonella*. Late genes required for filament formation (*flgL*, *fliC*, and *fliD*) are missing, agreeing with observations of the flagellar basal structures by electron microscopy (below). Although *flgK* belongs to the late genes, it exists at the end of an operon that encodes the basal body structure (rod, hook, PL ring, and HAP1). The genes on the chromosome are complete to form the basal structure observed.

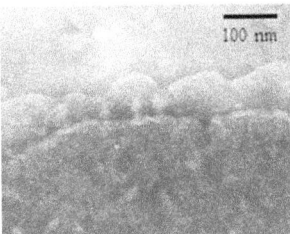

100 nm

The predicted isoelectric point (pI) of most of the *Buchnera* flagellar genes are **basic between pI 8.6–10.6**, whereas those of *Salmonella* flagellar genes are acidic between pI 4.6–6.4. *Buchnera* cells have to take in many nutrients from the host cells, but do not have conventional transport systems.

Urgent proposal for an interesting project

Circumstantial evidence supports an idea that the flagellar basal body might be used for taking in nutrients from the host. The secretion gate, gatekeepers, and Type III ATPase exist. Do they work in constructing the incomplete flagellar structures? Why are hundreds of them necessary?

Strain was provided by Masae Morioka of Tokyo University, Tokyo, Japan, as of 2006.

CHAPTER 32

Methanococcus voltae — Archaeal Flagella or Archaellum

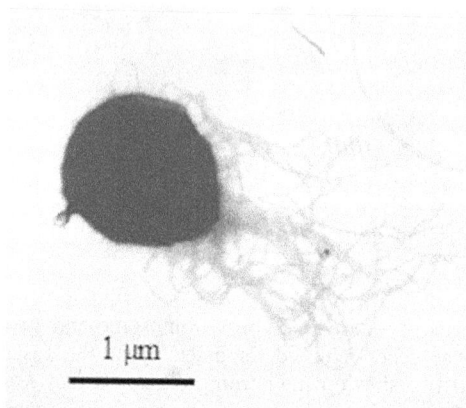

1 μm

Phylum	Euryarchaeota
Class	Methanococci
Order	Methanococcales
Family	Methanococcaceae
Genus	Methanococcus
Species	*M. voltae*

Methanococcus voltae is a Gram-negative coccoid-shaped archaeum. However, Gram reaction is pretty much meaningless for *Methanococcus* because the membrane constituents are different from those of eubacteria. This species is heterotrophic, **strictly anaerobic**, and mesophilic, and will **produce methane from CO_2/H_2** or formate. Each cell possesses more than 70 flagella on one side of the round cell. As is typical of archaeal flagella, *M. voltae* flagella are composed of multiple flagellins (see Topic: Multi-flagellin).

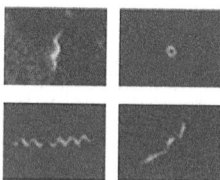

Filaments of *M. voltae* PS strain show **polymorphic transition** under various pH conditions (see Appendix). At neutral pH, filaments are Coiled. Between pH 3 and 5, they look Normal. At pHs of lower than 3, they become straight and then gradually melt. At alkaline pHs of between 11 and 13, they are Semi-Coiled.

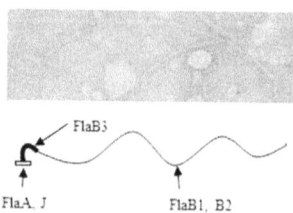

FlaB3

FlaA, J   FlaB1, B2

Purified flagella are shown to be composed of two major proteins, flagellins FlaB1 and FlaB2, with molecular masses corresponding to 33 and 31 kDa, respectively. FlaB3 is one of the minor components. The ratio of each flagellin in the filament is A:B1:B2:B3 = 0.1:10:10:1. Length of the hook portion occupies 5% of the filament length. From these facts, FlaB3 may form the hook portion.

Methanococcus voltae A3 genome (1,936,387 bp/ 1,717 genes)	NC_014222 CP002057

1,936/0 (Kbp)

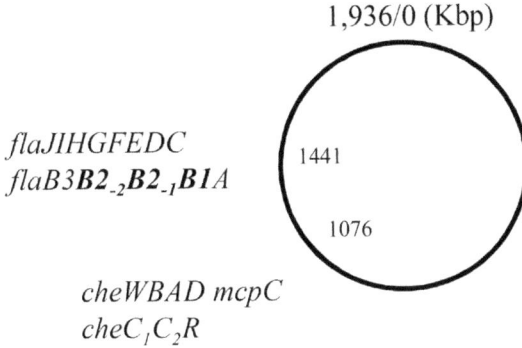

flaJIHGFEDC
flaB3B2$_{-2}$B2$_{-1}$B1A

cheWBAD mcpC
cheC$_1$C$_2$R

The genome of *Methanococcus voltae* A3 strain consists of one 1.9-Mbp chromosome. Flagellar genes are clustered at 1,441 Kbp on the chromosome. There are **five flagellin genes** (*flaA, flaB1,* two *flaB2,* and *flaB3*) at 1,441 bp, and the downstream co-transcribed flagellar accessory genes (*flaCDEFGHIJ*). The two flagellins of *flaB2* (*flaB2$_{-1}$* and *flaB2$_{-2}$*) are 75% homologous.

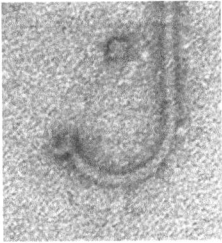

The proximal ends of the flagella show a hook and basal structure that looks like that of Gram-positives.

Despite many similarities to the flagellum of eubacteria (see Appendix), *M. voltae* flagellum has several properties that are different from those of the eubacterial flagellum. In the *M. voltae* flagellum:

1. Flagellins have **signal peptides**.
2. Flagellins contain **cysteines**, which do not exist in the eubacterial flagellin.
3. The number of flagellar genes is too few to construct the flagellum corresponding to that of eubacteria.

These facts strongly suggest that *M. voltae* flagellum might be different from the flagellum of eubacteria. It has been proposed to call the archaeal flagellum by a new name, **Archaellum**, to distinguish between the two.

Strains were provided by Ken Jarrell of The Queens University, Kingston, Canada. The project was carried out by Yoshika Nosaka, Shino Mizuno, and Kaoru Uchida.

CHAPTER 33

Myxococcus xanthus — To be Social or to be Adventurous

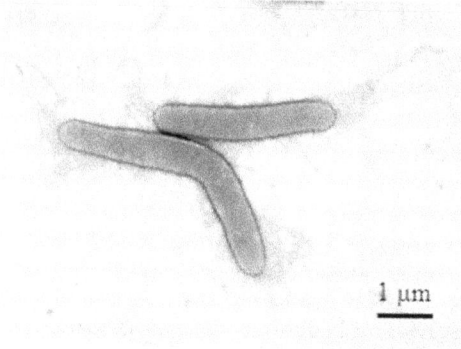

Phylum	Proteobacteria
Class	Deltaproteobacteria
Order	Myxococcales
Family	Myxococcaceae
Genus	Myxococcus
Species	*M. xanthus*

Myxococcus xanthus is a Gram-negative bacterium commonly found in soil that shows a social behavior: the cells move in a coordinated fashion and form a biofilm. *M. xanthus* cells **do not have flagella, but rather have type IV pili** and **glide on solid surfaces**. However, the pilus system may work by the same physics principle as the flagellum system; both have Mot or **Mot homologs powered by proton motive force**. The gliding motility is regulated by two motility systems: the A (adventurous) and the S (social) motility systems. A motility is a free movement of a single cell, while S motility is a coordinated movement of large cell groups. Compare with the social motility by flagella in *Paenibacillus* (see Chapter 17).

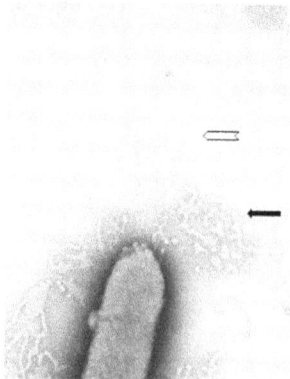

M. xanthus cells have pili (empty arrow) and extracellular matrix (filled arrow) on the leading pole. The pilus is a contractible straight filament with a diameter of 5–7 nm. In contrast, the extracellular matrix is a thick "flaccid" filament with a diameter of 50 nm.

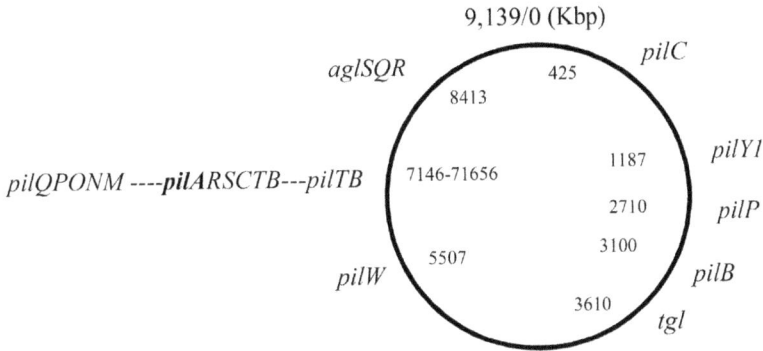

Myxococcus xanthus DK 1622 genome
(9,139,763 bp/ 7,316 genes)

NC_008095

9,139/0 (Kbp)

aglSQR

pilC

425

8413

1187   *pilY1*

pilQPONM ----pilARSCTB---pilTB   7146-71656

2710   *pilP*

3100

5507

pilW   3610   *pilB*

tgl

The genome of the *Myxococcus xanthus* DK 1622 strain consists of a single 9.1-Mbp chromosome. There is no flagellar gene. Instead, there are genes for assembly of type IV pilus, which is the "S-motility" organ of this gliding bacterium.

PilA is the major subunit of the pilus filament. Tgl is a lipoprotein and forms a transport system in the outer membrane together with PilQ secretin. AglR is a MotA homolog, and AglQ and AglS are MotB homologs, which lack the C-terminal peptidoglycan binding motif.

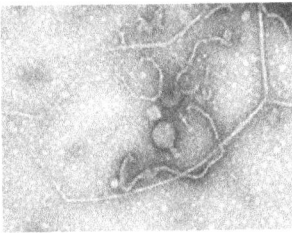

The type IV pilus is composed of several structural parts: adhesin, chaperone, usher, and filament. A crude preparation of pili tends to aggregate through its ends, suggesting its stickiness.

An osmotically-shocked cell reveals how pili originate from membranes. The basal structure in the outer and inner membranes is too tiny to be visible.

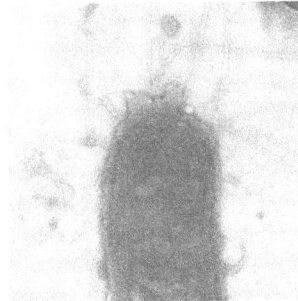

Strains were provided by David Zusman of the University of California at Berkeley, Berkeley, USA, and Zhaomin Yang of Virginia Polytechnic Institute and State University, Blacksburg, USA.

CHAPTER 34

Saprospira grandis — A Grand Predator on the Seashore

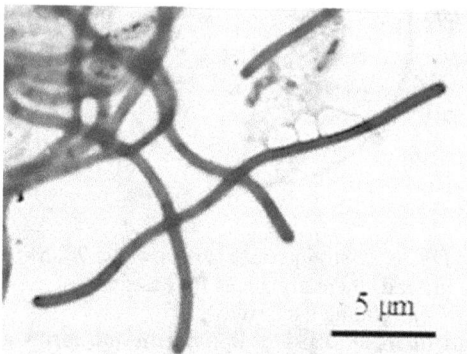

Phylum	Bacteroidetes
Class	Sphingobacteriia
Order	Sphingobacteriale
Family	Saprospiraceae
Genus	Saprospira
Species	*S. grandis*

Saprospira grandis is a Gram-negative, filamentous, marine bacterium. This is the only example of Sphingobacteria in this book. *S. grandis* cells glide over the solid surface. In sea water, they catch other bacteria on their cell surface by "ixotrophy." Ixotrophy is a kind of "fly-paper" system for catching the proteinaceous prey that it needs for its nutrition.

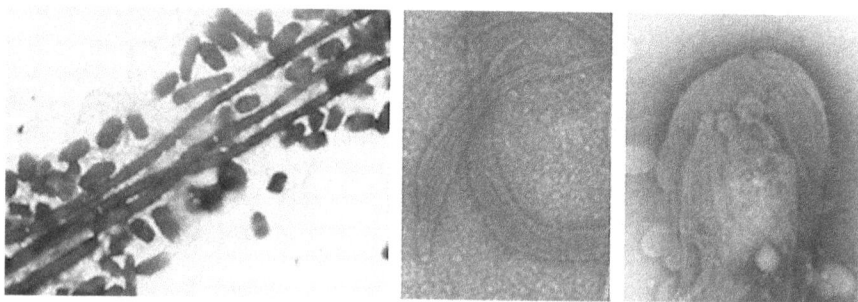

S. grandis cells catch flagellated *E. coli* cells more efficiently than aflagellated ones (left). The prey–predator aggregates become large enough to make a ball, which is lysed overnight. On the trace of the gliding path, numerous filamentous aggregates are observed (middle). These filaments are fragile and dissolve in water. When fixed with glutaraldehyde, a bundle of filaments can be observed to cover the cell surface (right).

Saprospira grandis Lewin genome (4,400,185 bp/ 4,251 genes)	NC_016940

4,345/0 (Kbp)

gldG
gldM
gldJ
gldN'
gldH

4293
3881
3105
2632
2248

778
1421
1599
1605
1695   2192

gldF
gldC
gldNML
gldK

gldDE - A   gldB

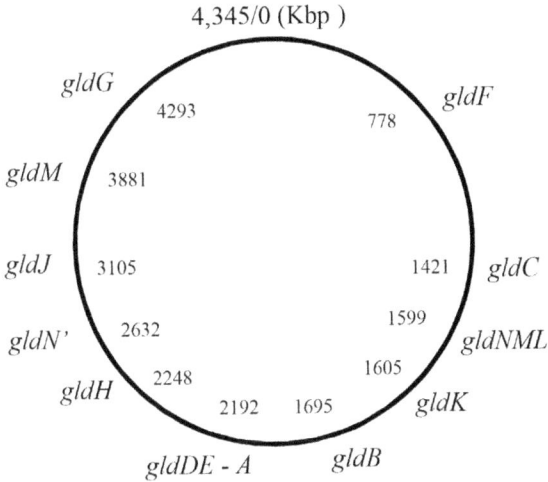

The genome of the *Saprospira grandis* Lewin strain consists of one 4,345,237 bp chromosome and a small plasmid (54,948 bp). Gliding motility (*gld*) genes are scattered all over the chromosome. Whether the gene products assemble into the filaments mentioned above is not known. There are a few genes for type IV pilus, but knowledge is still incomplete.

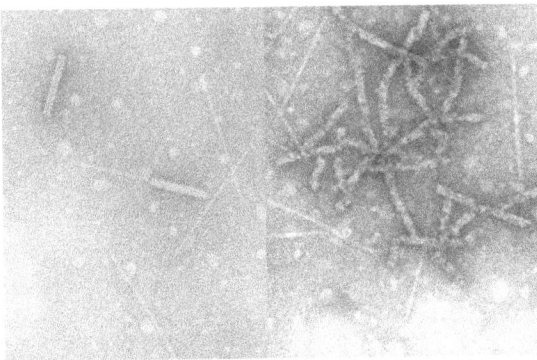

In the cytoplasm of the *Saprospira grandis* cell, there are numerous rod-shaped particles called rhapidosomes, which are composed of protein and RNA. A thin filament penetrates the center of a rod (left). The rod eventually disassembles into flexible filaments (right). Rhapidosome belongs to bacteriocins that kill bacteria.

Strains were provided by Ralph Lewin of Scripps Institute of Oceanography, San Diego, USA, as of 1995.
The project was carried out by Hiroyuki Mori and Masaomi Kanbe.

CHAPTER 35

Shigella flexneri — Flagellaless *E. coli*

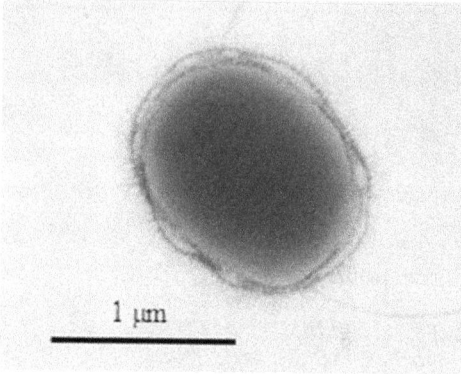

Phylum	Proteobacteria
Class	Gammaproteobacteria
Order	Enterobacteriales
Family	Enterobacteriaceae
Genus	Shigella
Species	*S. flexneri*

Shigella flexneri is a Gram-negative, non-spore forming, rod-shaped facultative anaerobe, which is physiologically similar to *Escherichia coli*. Although it is closely related to *E. coli*, *S. flexneri* can be differentiated because it lacks flagella and is thus nonmotile. Instead, *S. flexneri* cells possess numerous pili covering the whole cell surface.

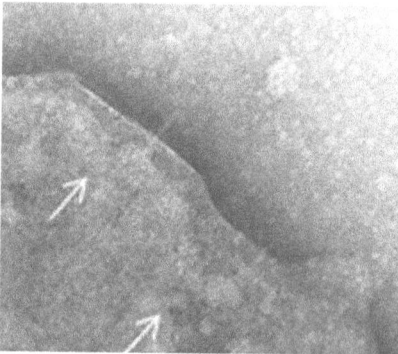

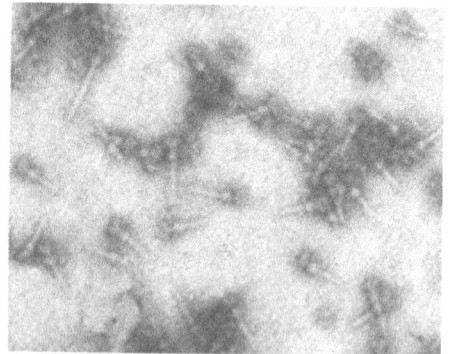

In general, *Shigella* species are human pathogens, causing severe gastroenteritis (bacillary dysentery) and resulting in over 1 million deaths a year. *S. flexneri* causes infection via a Type III secretion system. The secretion system produces numerous needles[1] (arrows) that inject the effector proteins (Ipa) into epithelial cells (see Topic: Flagella & Pathogenicity).

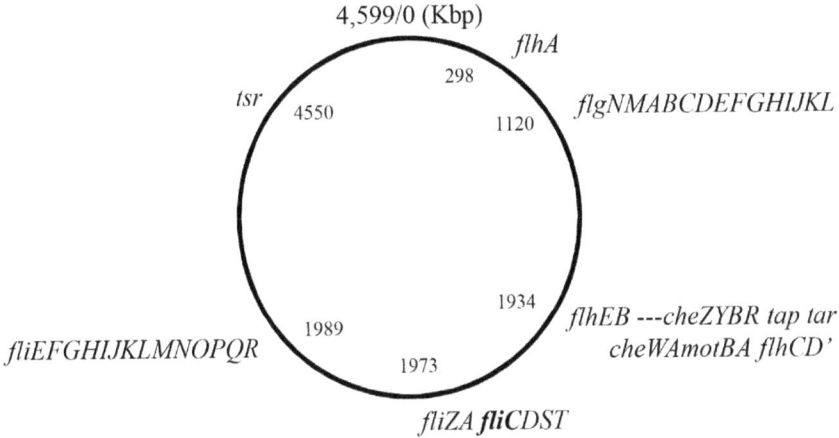

The genome of the *S. flexneri* 2457T strain consists of a single 4.5-Mbp chromosome. Almost all genes necessary for flagellar assembly are arranged in the same order as that in *Salmonella*. However, one of the master genes *flhD'* is truncated in half, and expression of the flagellar genes does not occur.[2,3] In a pathogenic *E. coli*, there is a **whole set of flagellar genes that is cryptic** as in *Shigella* (see Chapter 9, Flag-2).

The flagellin and hook protein predicted from *Shigella fliC* and *flgE* are highly homologous to those of *Salmonella/E. coli*. Whether *Shigella* cells can swim under certain conditions or not is a historical controversy.[4] Kiyoshi Shiga, who discovered the organism in 1897, claimed that it was motile.

> Urgent proposal for an interesting project

Replace the truncated *flhD'* with the intact *flhD* and see if the cell produces flagella or not. What do the flagella hidden for centuries look like?

Strains were provided by Chihiro Sasakawa of the University of Tokyo, Tokyo, Japan.
The project was carried out by Kaoru Komoriya.

APPENDIX

1 FLAGELLAR FAMILY

(A) Helical Filament

A flagellar filament has a helical shape. There are theoretically two types of helices, right-handed and left-handed. In nature, *Salmonella* has a left-handed filament, and *Caulobacter crescentus* has a right-handed filament under the right physiological conditions. It should be noted that shapes of these two helices are not mirror images of each other: that is, the pitch and diameter of each helix are not the same. In *Salmonella*, a filament is composed of a single kind of flagellin, but in many other species it is composed of more than two kinds of flagellins homologous to each other. As mentioned in Topic 9: Multiple flagellins, each flagellin may occupy different locations in a filament. However, the meaning of multiple flagellins is not clear. I will discuss the shapes and properties of filaments composed of a single type of flagellin.

There are several filament shapes, and it will be convenient to use the conventional names of typical shapes found in *Salmonella*: Normal (left-handed), Curly (right-handed), Coiled (left-handed), Semi-coiled (right-handed), and Straight.[1] Note that the names of filament shapes are indicated by adjectives starting with capital letters. The helical parameters of these helices are discrete and distinguishable from one another. A filament transforms its shape into several distinguishable helical shapes (polymorphs) under various environmental conditions. In *Salmonella*, as the pH value of the solution decreases, filaments change in form from Normal to Semi-coiled, to Coiled, to Curly, and then to Straight.[2] Straight filaments briefly appear just before disassembly begins. A helix is uniquely defined by three parameters: the pitch (P), the helix diameter (D), and the handedness. The helical parameters of typical polymorphs are shown in the following table.

Table A.1 Helical Parameters of Polymorphs of *Salmonella* Flagella			
Shape	Handedness	Pitch (μm)	Diameter (μm)
Normal	Left	2.55	0.60
Coiled	Left	$\fallingdotseq 0$	1.00
Semi-coiled	Right	1.29	0.50
Curly I	Right	1.20	0.20
Curly II	Right	1.00	0.15

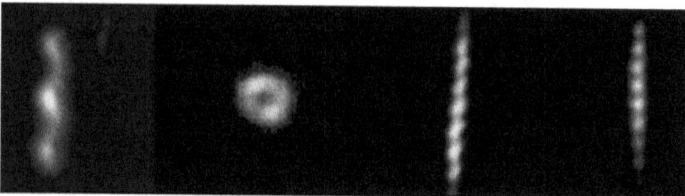

(from left to right, Normal, Coiled, Curly I, and Curly II).

The microscopic images of polymorphs of *Salmonella* flagellum are shown in the image on page 89. Filament shapes were observed by dark-field microscopy. The top surface of the filament was focused to show handedness. Normal is left-handed and Curly is right-handed.

(B) Polymorphism

A flagellar filament can switch between a set of helical shapes under appropriate conditions; not only helical pitch, but also helical handedness is changeable. The transformation of shapes can be induced by physical perturbation (torque, temperature, pH, salt concentration of medium).[3] Genetic changes, such as point mutations in the flagellin gene, also result in transformation of helices,[4] but some mutant filaments, such as straight filaments, are too stiff to transform into other helices.

This phenomenon, called "polymorphism," of the filament is a visible example of conformational changes in proteins, and therefore has evoked an idea of a functional role of the filament in motility. However, polymorphism of the filament by itself does not cause the rotary motion. Flagella are passive in terms of force generation. Polymorphism of filaments is observed to occur naturally on actively motile cells with peritrichous flagella. The helical transformation is necessary for untangling a jammed bundle of tangled filaments. When Normal-type filaments in a jammed bundle are transformed into Curly-type filaments, knots of tangled filaments run toward the free end of each flagellum to untangle the jammed bundle.[5]

(C) Calladine Model

Models that explain the polymorphism characteristic were first introduced by Sho Asakura[6] in 1970, and theoretically strengthened by Chris R. Calladine[7] in 1978. Twisting and bending a cylindrical rod gives rise to a helix. Models predict 12 shapes, and 8 of them have been found in existing filaments: Straight with a left-handed twist, fl, Normal, Coiled, Semi-Coiled, Curly I, Curly II, and Straight with a right-handed twist. Only a small energy barrier seems to occur between two neighboring shapes.[8] Polymorphic transition occurs from one shape to its neighbors; for example, in a transition from Normal to Curly I, the filament briefly takes on Coiled and Semi-Coiled forms.

(D) Flagellar Family

The polymorphs of a kind of filament are regarded as one flagellar family. In 2005, a marine bacteria *Idiomarina loihiensis* turned out to be different from the conventional flagellar family of *Salmonella*. The pitch and diameter of *I. loihiensis* filaments are the same as *Salmonella* Curly filaments, but are left-handed.[9] To explain this exceptional filament, it was necessary to analyze helical parameters of filaments of all species available in our lab. We first measured the pitch (P) and diameter (D) of each filament and its polymorphs. In order to summarize the results in a simple way, we did not use the Calladine plot, but developed a new plot: the pitch–diameter (P–πD) plot.[10] This is simply because the curvature and twist of a filament used in the Calladine model are calculated from (P, D) and are imaginary and immeasurable. On the other hand, the (P–πD) plot is simpler and more practical to use.

If the handedness was expressed as + (right handed) or − (left handed) of the pitch value, helices can be plotted on the (P–πD) plane. We learned that polymorphs of a type of flagellum stay on a circle in the (P–πD) plot, indicating that they correspond to one family in the Calladine model.

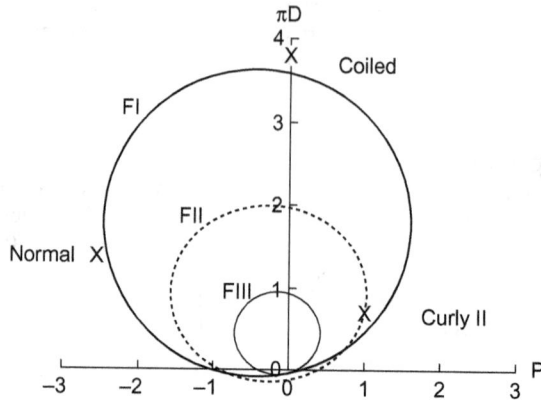

By analyzing flagellar helical parameters of numerous species, we found that flagella polymorphs of numerous species can be grouped into three families:

Family I: *S. typhimurium, E. coli, B. subtilis, E. casseliflavus, P. carotovorum, P. mirabilis, Y. enterocolitica;*

Family II: *I. loihiensis, A. brasilense Pof, B. japonicum Thick, P. aeruginosa, P. syringae, V. parahaemolyticus Pof, X. axonopodis;*

Family III: *A. brasilense Laf, B. japonica Thin, V. parahaemolyticus Laf.*

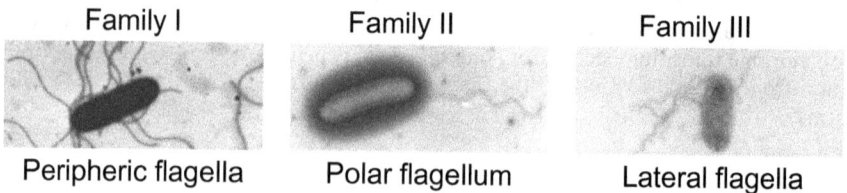

Family I	Family II	Family III
Peripheric flagella	Polar flagellum	Lateral flagella

Roughly speaking, a Family I filament has a large pitch and diameter, a Family II filament has a medium pitch and diameter, and a Family III filament has a small pitch and diameter. Coincidentally, Family I filaments are all from peritrichous flagella, Family II filaments are from polar flagellum, and Family III filaments are from lateral flagella. These facts tend to push us toward the idea that flagellar shape is determined by the position where the flagellum grows. However, as seen in *B. japonicum* (Chapter 7), both Family I flagella and Family II flagella grow from the subpolar region. In a *flhF* mutant of *V. parahaemolyticus*, the polar flagella become peritrichous flagella without changing helical parameters.[11] Thus it is likely that flagellar shape is determined independently from the position on a cell (see Topic: Flagellar position and shape).

There is one group of flagella that do not belong to any of the three families. *S. ruminantium* flagella show three different helical shapes by polymorphic transition: Normal, Coiled, and Curly, with the pitch and diameter larger than the counterparts of *S. typhimurium*. In the P–πD plot, these helices do not match on a closed circle.

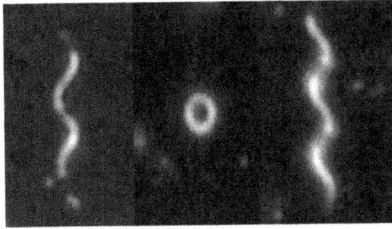

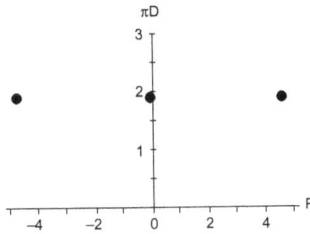

Therefore, we put them in the Exceptions of the flagellar family, to which flagella of *C. crescentus*, *R. sphaeroides*, *R. lupini*, and *S. meliloti* (all are alphaproteobacteria) also belong.

Family IV   *C. crescentus, R. sphaeroides, R. lupini, S. meliloti, S. ruminantium.*
(Exceptions)

C. crescentus, *S. meliloti*, and *S. ruminantium* filaments are composed of multiple flagellins, and thus their polymorphs are unpredictable. On the other hand, the *R. sphaeroides* filament is composed of a single flagellin. This is another mystery concerning helicity of flagellar filaments.

Calladine's mechanical model gives insight into the molecular mechanism of polymorphism. It predicts 12 polymorphs (including two straight filaments), and explains how those discrete polymorphs change one by one. This number of polymorphs is based on the assumption that the number of protofilaments in a filament is 11. However, it has been shown that the polar flagellum of *Campylobacter jejuni* has seven protofilaments.[12] It is tempting to conclude that the shape of flagellar families may be determined by the number of protofilaments, but this idea has to be tested with many more species.

2 METHODS FOR PURIFICATION OF FLAGELLA

There are several methods for purifying flagella, depending on which structure and which component proteins you want to analyze: flagellar filament or the basal structure. Purified filaments will give a single band (or several bands) of flagellin in SDS PAGE. Purified HBB will give several bands of protein components of the basal body. Using flow charts, I will describe the protocols used daily in our lab.

(A) Protocol I: Isolation and Purification of Filaments

The filament, the major part of the flagellum, is exposed to outside the cell and, thus, can be detached from the cell body by physical forces externally applied. Filaments thus isolated retain the hook attached at the proximal end. For the purpose of analyzing flagellin, this method is quick and simple to achieve (see the photograph next page).

Overnight culture
↓ For species that lose flagella after a prolonged incubation, harvest earlier.

Cell harvest by low-speed centrifugation
↓ Hard pellets are better to efficiently remove flagella.

Suspend cells in Phosphate-buffered saline (PBS)
↓ Make a condensed cell suspension to apply an efficient shearing force to the flagella. Phosphate buffer can be used, if cells of interest do not lyse in the absence of salts.

Passing through a hypodermic needle
↓ Use G25 or larger needles. After passing cells through the needle 10 times, observe the cells by microscopy, and, if most cells are still motile, repeat the procedure.

Remove cells by low-speed centrifugation
↓ Dilute the suspension with PBS so that detached filaments will not make aggregates. If cells remain in the supernatant as observed by microscopy, repeat centrifugation to remove cells completely.

Collect detached filaments by ultracentrifugation
↓ When treating a big batch, add PEG to the supernatants and collect aggregated flagella by low-speed centrifugation. Filaments alone make a translucent pellet. If the center of the pellet is pale brown with cell debris, collect only transparent part.

Suspend the filaments in the pellets with PBS
↓ Add PBS to cover the pellets and leave them standing in a cold temperature overnight. A hastily created suspension of pellets with a pipette will result in broken filaments.

Recover purified filaments in a tube and store in cold until use

The filaments isolated by this protocol have the hook and the distal rod attached at one end, indicating that the mechanically weakest part of the flagellum is at the midpoint of the rod.[1]

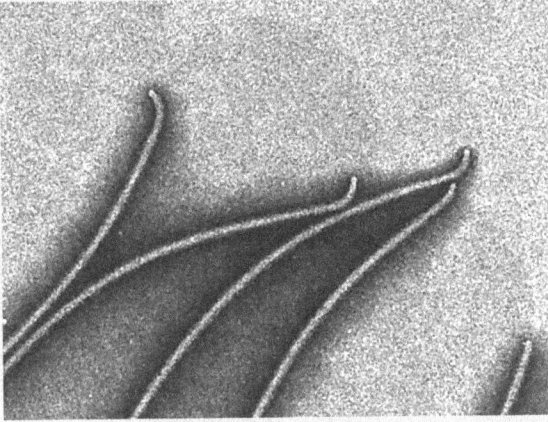

(B) Protocol II: Isolation and Purification of Intact Flagella

Intact flagellum is the filament that retains the hook–basal body attached at the proximal end. In order to isolate intact flagella, mild methods are required to prevent filaments from falling off the cell body. The most efficient way is for cells to be lysed using a combination of lysozyme and a nonionic detergent Triton X-100.[2]

Cell culture

↓ For a big batch of culture (more than 500 ml), aeration is one of the most important factors for vigorous growth and for preparation of cleaner samples. Use a large flask to give the culture a large surface area. Grow cells till the late log phase, but do not overgrow cells into the stationary phase.

Cell harvest by low-speed centrifugation

↓ Make soft pellets so that pellets can be completely dissolved in sucrose solution at the next step.

Suspend cells in sucrose solution

↓ The sucrose solution contains 0.5 M sucrose, 0.15 M Trizma base, pH not adjusted. This is important, because EDTA works better at alkaline pH. Add 1 mM protease inhibitor for species that secrete proteases such as *Bacillus* genus. Cells are slowly homogenized using a spatula and a wide-bore pipette. Treat them gently to avoid shearing off flagella from cells.

Add lysozyme solution (2 mg/ml in distilled water) at a final conc. 0.1 mg/ml.
↓ Place the pipette tip at the bottom of the cell suspension and add solution very slowly.

Add EDTA (final 10 mM)
↓ Place the pipette as mentioned above. For Gram-positive species, this step is skipped. After addition of EDTA, the beaker should be taken out of ice and placed in room temperature for lysozyme to work well. Check sphaeroplast formation (cells become round) by microscopy. The order of adding "lysozyme first and then EDTA" is important. "EDTA first and then lysozyme" does not work well.

Add Triton X-100
↓ Add 10% Triton X100 to a final conc. 1% to the stirring suspension at once from the above. The cloudy solution immediately becomes translucent and the center of the solution rises because of thick DNA. Viscosity is decreased by stirring at room temperature. Viscosity decreasing is a sign that endogenous DNases have degraded cellular DNA. After 30 min, if the solution does not become clear, stop experiments. Something has gone wrong.

Add Mg for DNA digestion by endogenous DNase
↓ For *Salmonella*, endogenous DNase digests DNA in 30 min without Mg addition. However, most Gram-negative species do require Mg for DNase to work. For some species, addition of Mg is not sufficient. In that case, add a tiny amount of powdered DNase. Judge the complete digestion of DNA using a pipette; if DNA remains, drops of lysate tend to stay at the tip. For Gram-positive species, this step is skipped.

Add EDTA
↓ To prevent reaggregation of cell membranes and walls by excess Mg ions, add the same amounts of EDTA as Mg.

Adjust pH of the solution at 10
↓ Add drops of 1N NaOH solution to suspension, constantly stirring. Frequently check the pH of the solution with a slip of pH indicator paper. At around 10, the solution becomes translucent. EDTA at alkaline pH seems to help dissolving the outer membrane vesicles.

Remove unlysed cells by low-speed centrifugation
↓ Spin down unlysed cells and aggregates of outer membranes. Repeat twice.

Collect flagella in the supernatant by ultracentrifugation
↓ The cleared supernatant is subjected to ultracentrifugation (75,000 × g, 60 min, in polyallomer tubes) and the pellets are suspended in alkaline solution (0.1 M KCl-KOH, 0.5 M sucrose, 0.1% Triton X-100 [pH 11]), and recentrifuged.

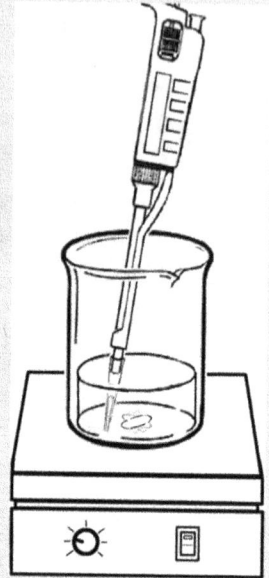

Resuspend the intact flagella in the pellets with TET
↓ Add TET buffer (10 mm Tris-HCl, 5 mM EDTA, 0.1% Triton X-100 [pH 8.0]) to cover the pellets and leave them standing in a cold temperature overnight.

Recover purified intact flagella and store in cold until use

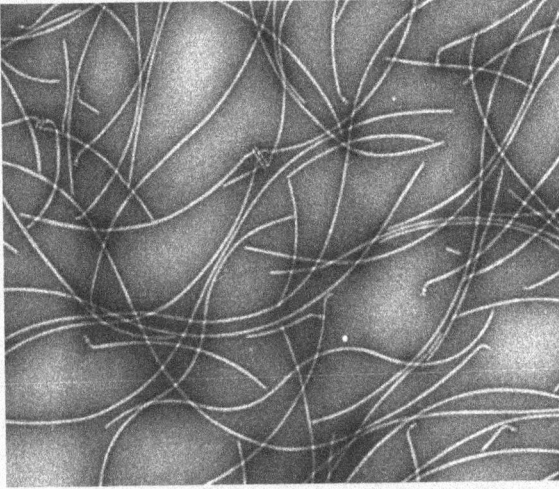

[Option] Protocol for CsCl density gradient centrifugation
When the purity of the intact flagella is unsatisfactory, you may further purify your samples using CsCl density gradient centrifugation. This step will separate flagella from membranous materials.

Dilute sample containing intact flagella with TET buffer
↓ The solution has to be thin enough so that flagella and membranes migrate without interfering with each other.

Add powder of CsCl into sample solution
↓ For Salmonella flagella, 33 (w/v)% CsCl is good for making a gradient in which the flagella band comes in the middle of the tube. Flagella from other species may have a different buoyant density in the CsCl solution. Try several concentrations between 30 and 40 (w/v)% CsCl.

Ultracentrifugation in a swing rotor (SW41Ti buckets, 68,000 × g, 16 h, 20°C).
↓ Set the temperature of the centrifugation at room temperature so that centrifuge tubes do not get condensation on the wall when they are taken out
from the buckets. Flagella should form a pale blue band near the midpoint
of the tube, membrane fragments should form a diffuse band above, and
debris should collect at the bottom. The flagellar fraction can be collected
with a Pasteur pipette.

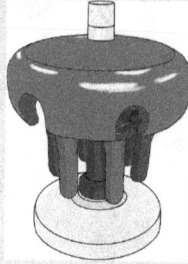

Dialyze the sample against TET buffer
Recover flagella solution and store in cold until use

(C) Protocol III: Preparation of HBB

Flagellar filament makes up 99% of the flagellum, judging from its size (10 micron filament vs. 100 nm HBB). To analyze HBB, filaments have to be completely removed.
There are only two ways to remove filament from the intact flagella: acidic pH or heat.
I will show the acid treatment below. The flagellar filaments are dissociated into monomeric flagellin in acidic solution, leaving the hook–basal bodies intact.

Dilute the flagella sample with distilled water and centrifuge
↓ The flagellar filaments are dissociated into monomeric flagellin by
Suspending the pellets in acidic buffer (50 mM glycine-hydrochloride, 1 mM
EDTA, 0.1% Triton X-100 [pH2.5]).
↓ Dissolve the pellets in a large amount of acidic buffer so that monomeric
flagellin does not repolymerize. Let the mixture suspend for half an hour
with a gentle shake.

Collect HBB complexes by ultracentrifugation(100,000 × g, 60 min).
↓ Residual flagellin from the supernatant should be carefully wiped from the
tube wall.
Suspend HBBs in TET buffer for EM or in SDS sample buffer for SDS-PAGE

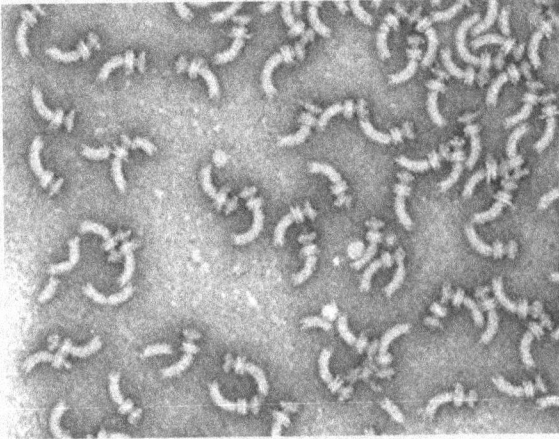

(D) Protocol IV: Preparation of Osmotically-Shocked Cells[3]

Cell culture
↓ Grow cells till the mid-log phase, but do not overgrow cells into the stationary phase.

Cell harvest by low-speed centrifugation
↓ Make soft pellets so that pellets can be completely dissolved in sucrose solution at the next step.

Suspend cells in sucrose solution
↓ The sucrose solution contains 0.5 M sucrose, 0.15 M Trizma base, pH not
adjusted. This is important, because EDTA works better at alkaline pHs.
Let mixture sit at room temperature for 10 min.

Dilute cell suspension into a large amount of cold water including 5 mM EDTA.
↓ This step has to be quick. Dump the mixture into stirring water.

Remove unlysed cells by low-speed centrifugation
↓ Most cells will be pellets at the bottom of the tube.
Spin down lysed cells by high-speed centrifugation
↓ The amount of lysed cells can be very small depending on the growth phase and degree of resistance against osmotic shocks.
Suspend cell pellet in water or diluted buffer
Observation by EM

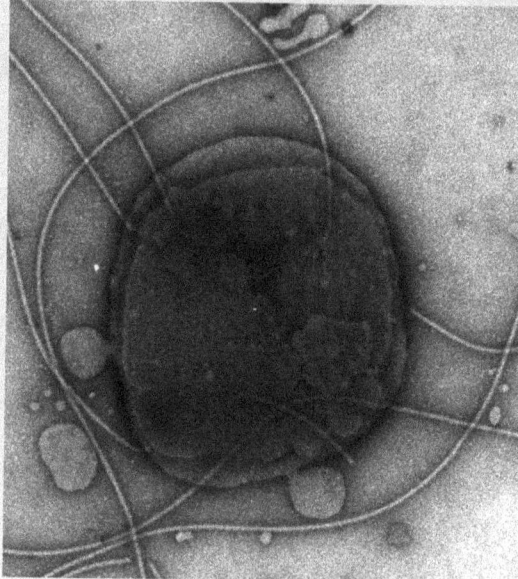

3 MICROSCOPIC METHODS FOR OBSERVATION OF FLAGELLA

(A) Phase-Contrast Microscopy

Flagellar filaments are too thin to observe under the phase-contrast microscope. However, a simple staining method of bacterial cells with crystal violet, safranin, or methylene blue allows us to visualize flagella under the phase-contrast microscope.

[Merits]   No sophisticated tools are required.
[Drawbacks]   Cells are dead. It takes time for the staining procedure.
[Best examples]   Leifson, E. (1969).[1]

(B) Dark-Field Microscopy

To illuminate filaments, a high-intensity lamp such as a mercury or a xenon lamp is used instead of the ordinary tungsten lamp. This also requires the oil immersion type of condenser lens and the objective lens with a large numerical aperture to obtain the dark background.

[Merits]   Best way to observe live cells with moving flagella. A single flagellum isolated from cells can be seen. Native helical parameters (helical pitch and diameter) are measured from recorded images.

[Drawbacks]   This kind of microscope is not commercially available. You have to assemble the microscope on your own.

[Best examples]   Macnab RM, Ornston MK. (1977).[2] Hotani, H. (1982).[3]

(C) Fluorescent Microscopy

Flagella of some species can be stained by fluorescent dye, such as Alexa Fluor 546 maleimide dye, and observed under the fluorescent microscope.

[Merits]   Fantastic images of flagella are obtained, because cells are alive and individual flagella can be seen.

[Drawbacks]   The microscope is expensive. Not all flagella are stained. Some cells lose their motility through the staining process.

[Best examples]   Turner L, Ryu WS, Berg HC. (2000),[4] Turner L, Stern AS, Berg HC. (2012).[5]

(D) Electron Microscopy

Negative staining is the best way to observe cells with flagella. But, there will be several possible difficulties for beginners when attempting the observation of flagella by electron microscopy. To avoid issues, pay careful attention to the next points: (1) Charge the grid surface before use. Carbon coated surface is hydrophobic. Glow discharging of grids is necessary for samples to stay and evenly scatter on the grid surface. (2) Put a drop of staining solution (PTA) on the grid first and then add an aliquot of sample in the drop and wait for a minute until the sample sits on the grid surface. Cells are still motile in the staining solution. (3) Survey a grid for spots properly stained (not too dark and not too light) at low magnification and go to higher magnifications at that spot. Observe samples at ×5,000 for cells and ×20,000 for flagella. Magnification higher than that will not necessarily give a higher resolution.

[Merits]   Simple to prepare specimens. Molecular details of the structures can be seen (see Introduction).

[Drawbacks]   Samples are dried. Maintenance of the machine is expensive.

[Best examples]   This book.

(E) Scanning Tunneling Microscopy (STM)/Atomic Force Microscopy (AFM)

AFM provides highly magnified images of samples in a solution. Samples have to stick on the fresh surface of a mica sheet.[6]

[Merits]   Wet samples can be observed. Images can have a high resolution at the atomic level.

[Drawbacks]   Filament is too large to observe. Machines are expensive, and it requires special techniques to operate the machine.

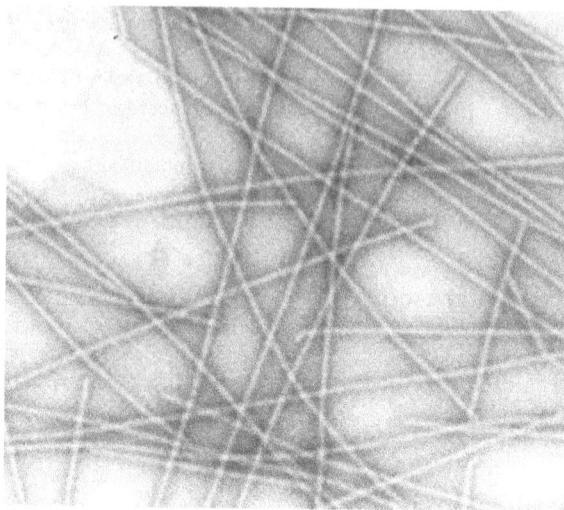

[Best examples]   Jonas K, Tomenius H, Kader A, Normark S, Römling U, Belova LM, et al. (2007),[6] Naya M, Mononobe S, Maheswari RM, Saiki T, Ohtsu M. (1996),[7] Naya, M, Micheletto R, Mononobe S, Uma Maheswari R, Ohtsu M. (1997).[8]

REFERENCES

INTRODUCTION

1. Leifson E. *Atlas of bacterial flagellation.* New York and London: Academic Press; 1969.
2. CDC. *Salmonella surveillance: Annual summary 2005.* Atlanta, Georgia: US Department of Health and Human Services, CDC; 2007.
3. DePamphilis ML, Adler J. Purification of intact flagella from Escherichia coli and Bacillus subtilis. *J Bacteriol* 1971;**105**:376–83.
4. DePamphilis ML, Adler J. Fine structure and isolation of the hook-basal body complex of flagella from Escherichia coli and Bacillus subtilis. *J Bacteriol* 1971;**105**:384–95.
5. DePamphilis ML, Adler J. Attachment of flagellar basal bodies to the cell envelope: specific attachment to the outer, lipopolysaccharide membrane and the cytoplasmic membrane. *J Bacteriol* 1971;**105**:396–407.
6. Aizawa S-I, Dean GE, Jones CJ, Macnab RM, Yamaguchi S. Purification and characterization of the flagellar hook-basal body complex of Salmonella typhimurium. *J Bacteriol* 1985;**161**:836–49.
7. Aizawa S-I. Flagellar assembly in *Salmonella typhimurium. Mol Microbiol* 1996;**19**:1–5.
8. Aizawa S-I. Flagella. In: Lederberg J, editor. *Encyclopedia of microbiology.* New York: Academic Press; 2000. p. 380–9.
9. Aizawa S-I. Flagella. In: Schaechter M, editor. *Encyclopedia of microbiology.* Oxford: Elsevier; 2009. p. 393–403.
10. Aizawa S-I. Mystery of FliK in length control of the Flagellar Hook. *J Bacteriol* 2012;**194**:4798–800.
11. Aizawa S-I. Rebuttal: flagellar hook length is controlled by a secreted molecular ruler. *J Bacteriol* 2012;**194**:4797.
12. Hughes KT. Flagellar hook length is controlled by a secreted molecular ruler. *J Bacteriol* 2012;**194**:4793–6.
13. Hughes KT. Rebuttal: mystery of FliK in length control of the flagellar hook. *J Bacteriol* 2012;**194**:4801.
14. Fahner KA, Block SM, Krishnaswamy S, Parkinson JP, Berg HC. A mutant hook-associated protein (HAP3) facilitates torsionally induced transformations of the flagellar filament of Escherichia coli. *J Mol Biol* 1994;**238**:173–86.
15. Khan IH, Reese TS, Khan S. The cytoplasmic component of the bacterial flagellar motor. *Proc Natl Acad Sci USA* 1992;**89**:5956–60.
16. Francis NR, Sosinsky GE, Thomas D, DeRosier DJ. Isolation, characterization and structure of bacterial flagellar motors containing the switch complex. *J Mol Biol* 1994;**235**:1261–70.
17. Katayama E, Shiraishi T, Oosawa K, Baba N, Aizawa S-I. Geometry of the flagellar motor in the cytoplasmic membrane of Salmonella typhimurium as determined by stereo-photogrammetry of quick-freeze deep-etch replica images. *J Mol Biol* 1996;**255**:458–75.
18. Minamino T, González-Pedrajo B, Kihara M, Namba K, Macnab RM. The ATPase FliI can interact with the type III flagellar protein export apparatus in the absence of its regulator, FliH. *J Bacteriol* 2003;**185**:3983–8.
19. Aldridge P, Karlinsey JE, Becker E, Chevance FF, Hughes KT. Flk prevents premature secretion of the anti-sigma factor FlgM into the periplasm. *Mol Microbiol* 2006;**60**:630–43.
20. Kubori T, Shimamoto N, Yamaguchi S, Namba K, Aizawa S-I. Morphological pathway of flagellar assembly in salmonella typhymurium. *J Mol Biol* 1992;**226**:433–46.
21. Ueno T, Oosawa K, Aizawa S. M ring, S ring and proximal rod of the flagellar basal body of Salmonella typhimurium are composed of subunits of a single protein, FliF. *J Mol Biol* 1992;**227**(3):672–7.
22. Schoenhals GJ, Macnab RM. Physiological and biochemical analyses of FlgH, a lipoprotein forming the outer membrane L ring of the flagellar basal body of Salmonella typhimurium. *J Bacteriol* 1996;**178**(14):4200–7.

23. Liu R, Ochman H. Stepwise formation of the bacterial flagellar system. *Proc Natl Acad Sci USA* 2007;**104**(17):7116–21.
24. Hueck CJ. Type III protein secretion systems in bacterial pathogens of animals and plants. *Microbiol & Mol Biol Rev* 1998;**62**:379–433.
25. Iino T, Komeda Y, Kutsukake K, Macnab RM, Matsumura P, Parkinson JS, et al. New unified nomenclature for the flagellar genes of Escherichia coli and Salmonella typhimurium. *Microbiol Rev* 1988;**52**:533–5.
26. Kaimer C, Berleman JE, Zusman DR. Chemosensory signaling controls motility and subcellular polarity in Myxococcus xanthus. *Curr Opin Microbiol* 2012;**15**(6):751–7.
27. Hazelbauer GL. Bacterial chemotaxis: the early years of molecular studies. *Annu Rev Microbiol* 2012;**66**:285–303.
28. Sourjik V, Wingreen NS. Responding to chemical gradients: bacterial chemotaxis. *Curr Opin Cell Biol* 2012;**24**(2):262–8.
29. Kutsukake K, Iino T. A trans-acting factor mediates inversion of a specific DNA segment in flagellar phase variation of Salmonella. *Nature* 1980;**284**(5755):479–81.
30. Silverman M, Simon M. Phase variation: genetic analysis of switching mutants. *Cell* 1980;**19**:845–54.
31. Kutsukake K. Hook-length control of the export-switching machinery involves a double-locked gate in Salmonella typhimurium. *J Bacteriol* 1997;**179**:1268–73.
32. Aldridge P, Karlinsey JE, Becker E, Chevance FF, Hughes KT. Flk prevents premature secretion of the anti-sigma factor FlgM into the periplasm. *Mol Microbiol* 2006;**60**(3):630–43.

CHAPTER EXAMPLES

1. Hamer R, Chen PY, Armitage JP, Reinert G, Deane CM. Deciphering chemotaxis pathways using cross species comparisons. *BMC Syst Biol* 2010;**4**:3.
2. Aizawa S-I. What is essential for flagellar assembly? In: Jarrell K, editor. *Pili and flagella: current research and future trends.* Horizonbook, Caister Academic Press; 2009. p. 91–8.
3. Kutsukake K, Ikebe T, Yamamoto S. Two novel regulatory genes, fliT and fliZ, in the flagellar regulon of Salmonella. *Genes Genet Syst* 1999;**74**(6):287–92.
4. Hung CC, Haines L, Altier C. The flagellar regulator fliT represses Salmonella pathogenicity island 1 through flhDC and fliZ. *PLoS One* 2012;**7**(3):e34220.
5. Barker CS, Meshcheryakova IV, Kostyukova AS, Samatey FA. FliO regulation of FliP in the formation of the Salmonella enterica flagellum. *PLoS Genet* 2010;**6**(9):e1001143.
6. Kusumoto A, Shinohara A, Terashima H, Kojima S, Yakushi T, Homma M. Collaboration of FlhF and FlhG to regulate polar-flagella number and localization in Vibrio alginolyticus. *Microbiology* 2008;**154**(Pt 5):1390–9.
7. Kutsukake K, Ohya Y, Iino T. Transcriptional analysis of the flagellar regulon of Salmonella typhimurium. *J Bacteriol* 1990;**172**(2):741–7.

CHAPTER 1. ACTINOPLANES MISSOURIENSIS

1. Uchida K, Jang M-S, Ohnishi Y, Horinouchi S, Hayakawa M, Fujita N, et al. Characterization of spore flagella in Actinoplanes missouriensis. *Appl Environ Microbiol* 2011;**77**:2559–62.
2. Yamamura H, Ohnishi Y, Ishikawa J, Ichikawa N, Ikeda H, Sekine M, et al. Complete genome sequence of the motile actinomycete Actinoplanes missouriensis 431(T) (=NBRC 102363(T)). *Stand Genomic Sci* 2012;**7**:294–303.

CHAPTER 2. ALIIVIBRIO FISHEREI

1. Millikan DS, Ruby EG. Vibrio fischeri flagellin A is essential for normal motility and for symbiotic competence during initial squid light organ colonization. *J Bacteriol* 2004;**186**:4315–25.
2. Ruby EG, Urbanowski M, Campbell J, Dunn A, Faini M, Gunsalus R, et al. Complete genome sequence of Vibrio fischeri: a symbiotic bacterium with pathogenic congeners. *Proc Natl Acad Sci USA* 2005;**102**:3004–9.
3. Urbanczyk H, Ast JC, Higgins MJ, Carson J, Dunlap PV. Reclassification of Vibrio fischeri, Vibrio logei, Vibrio salmonicida and Vibrio wodanis as Aliivibrio fischeri gen. nov., comb. nov.,

Aliivibrio logei comb. nov., Aliivibrio salmonicida comb. nov. and Aliivibrio wodanis comb. nov. *Int J Syst Evol Microbiol* 2007;**57**:2823–9.
4. Brennan CA, Mandel MJ, Gyllborg MC, Thomasgard KA, Ruby EG. Genetic determinants of swimming motility in the squid light-organ symbiont Vibrio fischeri. *Microbiol Open* 2013;**2**(4):576–94.

CHAPTER 3. AZOSPIRILLUM BRASILENSIS

1. Moens S, Michiels K, Keijers V, Van Leuven F, Vanderleyden J. Cloning, sequencing, and phenotypic analysis of laf1, encoding the flagellin of the lateral flagella of Azospirillum brasilense Sp7. *J Bacteriol* 1995;**177**(19):5419–26.
2. Scheludko AV, Katsy EI, Ostudin NA, Gringauz OK, Panasenko VI. Novel classes of Azospirillum brasilense mutants with defects in the assembly and functioning of polar and lateral flagella. *Mol Gen Mikrobiol Virusol* 1998;**4**:33–7.
3. Alexandre G, Greer SE, Zhulin IB. Energy taxis is the dominant behavior in Azospirillum brasilense. *J Bacteriol* 2000;**182**:6042–8.

CHAPTER 4. BACILLUS SUBTILIS

1. DePamphilis ML, Adler J. Fine structure and isolation of the hook-basal body complex of flagella from Escherichia coli and Bacillus subtilis. *J Bacteriol* 1971;**105**:384–95.
2. Kubori T, Okumura M, Kobayashi N, Nakamura D, Iwakura M, Aizawa S-I. Purification and characterization of the flagellar hook-basal body complex of *Bacillus subtilis*. *Mol Microbiol* 1997;**24**:399–410.
3. Aizawa S-I, Zhulin I, Márquez-Magaña L, Ordal GW. Chemotaxis and Motility in *Bacillus subtilis*. In: *Bacillus subtilis and its closest relatives: from genes to cells*. Washington, D.C: ASM press; 2002. p. 437–52.
4. Patrick JE, Kearns DB. Swarming motility and the control of master regulators of flagellar biosynthesis. *Mol Microbiol* 2012;**83**:14–23.
5. Guttenplan SB, Shaw S, Kearns DB. The cell biology of peritrichous flagella in Bacillus subtilis. *Mol Microbiol* 2013;**87**:211–29.
6. Ito M, Hicks DB, Henkin TM, Guffanti AA, Powers B, et al. MotPS is the stator-force generator for motility of alkaliphilic Bacillus and its homologue is a second functional Mot in Bacillus subtilis. *Mol Microbiol* 2004;**53**:1035–49.

CHAPTER 5. BDELLOVIBRIO BACTEROVORUS

1. Rendulic S, Jagtap P, Rosinus A, Eppinger M, Baar C, Lanz C, et al. A predator unmasked: life cycle of Bdellovibrio bacteriovorus from a genomic perspective. *Science* 2004;**303**(5658):689–92.
2. Lambert C, Evans KJ, Till R, Hobley L, Capeness M, Rendulic S. Characterizing the flagellar filament and the role of motility in bacterial prey-penetration by *Bdellovibrio* bacteriovorus. *Mol Microbiol* 2006;**60**:274–86.
3. Evans KJ, Lambert C, Sockett RE. Predation by Bdellovibrio bacteriovorus HD100 requires type IV pili. *J Bacteriol* 2007;**189**:4850–9.
4. Morehouse KA, Hobley L, Capeness M, Sockett RE. Three motAB stator gene products in bdellovibrio bacteriovorus contribute to motility of a single flagellum during predatory and prey-independent growth. *J Bacteriol* 2011;**193**:932–43.

CHAPTER 6. BORRELIA BURGDORFERI

1. Fraser CM, Casjens S, Huang WM, Sutton GG, Clayton R, Lathigra R, et al. Genomic sequence of a Lyme disease spirochaete, Borrelia burgdorferi. *Nature* 1997;**390**:580–6.
2. Charon NW, Cockburn A, Li C, Liu J, Miller KA, Miller MR, et al. The unique paradigm of spirochete motility and chemotaxis. *Annu Rev Microbiol* 2012;**66**:349–70.

CHAPTER 7. BRADYRHIZOBIUM JAPONICUM

1. Kanbe M, Yagasaki J, Zehner S, Göttfert M, Aizawa S-I. Characterization of two sets of sub-polar flagella in *Bradyrhizobium japonicum*. *J Bacteriol* 2007;**189**:1083–9.
2. Tsukui T, Eda S, Kaneko T, Sato S, Okazaki S, Kakizaki-Chiba K, et al. The type III secretion system of Bradyrhizobium japonicum USDA122 mediates symbiotic incompatibility with Rj2 soybean. *Appl Environ Microbiol* 2012;**79**:1048–51.

CHAPTER 8. CAULOBACTER CRESCENTUS

1. Paul R, Weiser S, Amiot NC, Chan C, Schirmer T, Giese B, et al. Cell cycle-dependent dynamic localization of a bacterial response regulator with a novel di-guanylate cyclase output domain. *Genes Dev* 2004;**18**(6):715–27.
2. Kanbe M, Shibata S, Jenal U, Aizawa S-I. Protease susceptibility of the Caulobacter crescentus flagellar Hook-Basal-Body; a possible mechanism of flagellar ejection during cell differentiation. *Microbiology* 2005;**151**:433–8.
3. Faulds-Pain A, Birchall C, Aldridge C, Smith WD, Grimaldi G, Nakamura S, et al. Flagellin redundancy in Caulobacter crescentus and its implications for flagellar filament assembly. *J Bacteriol* 2011;**193**(11):2695–707.

CHAPTER 9. ENTEROCOCCUS CASSELIFLAVUS

1. Kondoh M, Motonaga C, Okamori M, Hayashi A, Nohmi T, Shimada T, et al. Subacute toxicity study of ethanol-treated flagellate Enterococcus casseliflavus NF-1004 strain (NP-04) in rats. *Pharmacometrics* 2005;**68**(3/4):103–11.
2. Ferguson DM, Griffith JF, McGee CD, Weisberg SB, Hagedorn C. Comparison of enterococcus species diversity in marine water and wastewater using enterolert and EPA method 1600. *J Environ Public Health* 2013;**2013**:848049.

CHAPTER 10. ESCHERICHIA COLI

1. Block SM, Blair DF, Berg HC. Successive incorporation of force-generating units in the bacterial rotary motor. *Nature* 1984;**309**:470–2.
2. Blair DF, Berg HC. Restoration of torque in defective flagellar motors. *Science* 1988;**242**:1678–81.
3. Berg HC, Turner L. Torque generated by the flagellar motor of Escherichia coli. *BiophysJ* 1993;**65**:2201–16.
4. Berry RM, Berg HC. Absence of a barrier to backwards rotation of the bacterial flagellar motor demonstrated with optical tweezers. *Proc Natl Acad Sci USA* 1997;**94**:14433–7.
5. Lele PP, Branch RW, Nathan VS, Berg HC. Mechanism for adaptive remodeling of the bacterial flagellar switch. *Proc Natl Acad Sci USA* 2012;**109**(49):20018–22.
6. Adler J, Templeton B. The effect of environmental conditions on the motility of Escherichia coli. *J Gen Microbiol* 1967;**46**:175–84.
7. Inaba S, Hashimoto M, Jyot J, Aizawa S-I. Exchangeability of the flagellin (FliC) and the cap protein (FliD) among different species in flagellar assembly. *Biopolymers* 2013;**99**:63–72.
8. Morris RT, Drouin G. Ectopic gene conversions in bacterial genomes. *Genome* 2007;**50**(11):975–84.
9. Rode CK, Melkerson-Watson LJ, Johnson AT, Bloch CA. Type-specific contributions to chromosome size differences in Escherichia coli. *Infect Immun* 1999;**67**(1):230–6.
10. Daniell SJ, Takahashi N, Wilson R, Friedberg D, Rosenshine I, Booy FP, et al. The filamentous type III secretion translocation of enteropathogenic Escherichia coli. *Cell Microbiol* 2001;**3**:865–71.
11. Chaudhuri RR, Sebaihia M, Hobman JL, Webber MA, Leyton DL, Goldberg MD, et al. Complete genome sequence and comparative metabolic profiling of the prototypical enteroaggregative Escherichia coli strain 042. *PLoS One* 2010;**5**:e8801.
12. Ren CP, Beatson SA, Parkhill J, Pallen MJ. The Flag-2 locus, an ancestral gene cluster, is potentially associated with a novel flagellar system from Escherichia coli. *J Bacteriol* 2005;**187**(4):1430–40.

13. Makino S, Tobe T, Asakura H, Watarai M, Ikeda T, Takeshi K, et al. Distribution of the secondary type III secretion system locus found in enterohemorrhagic Escherichia coli O157:H7 isolates among Shiga toxin-producing E. coli strains. *J Clin Microbiol* 2003;**41**(6):2341–7.

CHAPTER 11. GEOBACILLUS KAUSTOPHILUS

1. Takami H, Takaki Y, Chee GJ, Nishi S, Shimamura S, Suzuki H, et al. Thermoadaptation trait revealed by the genome sequence of thermophilic Geobacillus kaustophilus. *Nucleic Acids Res* 2004;**32**:6292–303.
2. Yoshida K, Sanbongi A, Murakami A, Suzuki H, Takenaka S, Takami H. Three inositol dehydrogenases involved in utilization and interconversion of inositol stereoisomers in a thermophile, Geobacillus kaustophilus HTA426. *Microbiology* 2012;**158**(Pt 8):1942–52.

CHAPTER 12. GLUCONOBACTER OXYDANS

1. Prust C, Hoffmeister M, Liesegang H, Wiezer A, Fricke WF, Ehrenreich A, et al. Complete genome sequence of the acetic acid bacterium Gluconobacter oxydans. *Nat Biotechnol* 2005;**23**:195–200.
2. Malimas T, Yukphan P, Takahashi M, Muramatsu Y, Kaneyasu M, Potacharoen W, et al. Gluconobacter japonicus sp. nov., an acetic acid bacterium in the Alphaproteobacteria. *Int J Syst Evol Microbiol* 2009;**59**(Pt 3):466–71.

CHAPTER 13. HELICOBACTER PYLORI

1. Suerbaum S, Michetti P. *Helicobacter pylori* infection. *N Engl J Med* 2002;**347**:1175–86.
2. Josenhans C, Labigne A, Suerbaum S. Comparative ultrastructural and functional studies of *Helicobacter pylori* and *Helicobacter mustelae* flagellin mutants: both flagellin subunits, FlaA and FlaB, are necessary for full motility in *Helicobacter* species. *J Bacteriol* 1995;**177**:3010–20.
3. Suerbaum S, Josenhans C. The role of motility as a virulence factor in bacteria. *Int J Med Microbiol* 2002;**291**:605–14. [Review].
4. Tomb JF, White O, Kerlavage AR, Clayton RA, Sutton GG, Fleischmann RD, et al. The complete genome sequence of the gastric pathogen Helicobacter pylori. *Nature* 1997;**388**:539–47.
5. Niehus E, Gressmann H, Ye F, Schlapbach R, Dehio M, Dehio C, et al. Genome-wide analysis of transcriptional hierarchy and feedback regulation in the flagellar system of *Helicobacter pylori*. *Mol Microbiol* 2004;**52**:947–61.
6. Behrens W, Bönig T, Suerbaum S, Josenhans C. Genome sequence of Helicobacter pylori hpEurope strain N6. *J Bacteriol* 2012;**194**:3725–6.
7. Kostrzynska M, Betts JD, Austin JW, Trust TJ. Identification, characterization, and spatial localization of two flagellin species in Helicobacter pylori flagella. *J Bacteriol* 1991;**173**:937–46.

CHAPTER 14. IDIOMARINA LOICHENSIS

1. Shibata S, Alam M, Aizawa S-I. Flagella of the deep-sea bacteria Idiomarina loihiensis belong to a family different from Salmonella flagella. *J Mol Biol* 2005;**352**:510–6.
2. Hou S, Saw JH, Lee KS, Freitas TA, Belisle C, Kawarabayasi Y, et al. Genome sequence of the deep-sea gamma-proteobacterium *Idiomarina loihiensis* reveals amino acid fermentation as a source of carbon and energy. *Proc Natl Acad Sci USA* 2004;**101**:18036–41.

CHAPTER 15. LEGIONELLA PNEUMOPHILA

1. Glöckner G, Albert-Weissenberger C, Weinmann E, Jacobi S, Schunder E, Steinert M, et al. Identification and characterization of a new conjugation/type IVA secretion system (trb/tra) of Legionella pneumophila Corby localized on two mobile genomic islands. *Int J Med Microbiol* 2008;**298**:411–28.

2. Albert-Weissenberger C, Sahr T, Sismeiro O, Hacker J, Heuner K, Buchrieser C. Control of flagellar gene regulation in Legionella pneumophila and its relation to growth phase. *J Bacteriol* 2010;**192**:446–55.
3. Schroeder GN, Petty NK, Mousnier A, Harding CR, Vogrin AJ, Wee B, et al. Legionella pneumophila strain 130b possesses a unique combination of type IV secretion systems and novel Dot/Icm secretion system effector proteins. *J Bacteriol* 2010;**192**(22):6001–16.

CHAPTER 16. MAGNETOSPIRILLUM MAGNETOTACTICUM

1. Schultheiss D, Kube M, Schüler D. Inactivation of the flagellin gene flaA in Magnetospirillum gryphiswaldense results in nonmagnetotactic mutants lacking flagellar filaments. *Appl Environ Microbiol* 2004;**70**(6):3624–31.
2. Komeili A. Molecular mechanisms of compartmentalization and biomineralization in magnetotactic bacteria. *FEMS Microbiol Rev* 2012;**36**:232–55.
3. Ruan J, Kato T, Santini CL, Miyata T, Kawamoto A, Zhang WJ, et al. Architecture of a flagellar apparatus in the fast-swimming magnetotactic bacterium MO-1. *Proc Natl Acad Sci USA* 2012;**109**:20643–8.
4. Sakaguchi S, Taoka A, Fukumori Y. Analysis of magnetotactic behavior by swimming assay. *Biosci Biotechnol Biochem* 2013;**77**:940–7.

CHAPTER 17. PAENIBACILLUS ALVEI

1. Ingham CJ, Ben Jacob E. Swarming and complex pattern formation in Paenibacillus vortex studied by imaging and tracking cells. *BMC Microbiol* 2008;**8**:36.
2. Zarschler K, Janesch B, Zayni S, Schäffer C, Messner P. Construction of a gene knockout system for application in Paenibacillus alvei CCM 2051T, exemplified by the S-layer glycan biosynthesis initiation enzyme WsfP. *Appl Environ Microbiol* 2009;**75**:3077–85.

CHAPTER 18. PECTOBACTERIUM CAROTOVORUM

1. Chan YC, Wu HP, Chuang DY. Extracellular secretion of Carocin S1 in Pectobacterium carotovorum subsp. carotovorum occurs via the type III secretion system integral to the bacterial flagellum. *BMC Microbiol* 2009;**9**:181.
2. Matsumoto H, Muroi H, Umehara M, Yoshitake Y, Tsuyumu S. Peh production, flagellum synthesis, and virulence reduced in Erwinia carotovora subsp. carotovora by mutation in a homologue of cytR. *Mol Plant Microbe Interact* 2003;**16**:389–97.

CHAPTER 19. PSEUDOMONAS AERUGINOSA

1. Stover CK, Stover CK, Pham XQ, Erwin AL, Mizoguchi SD, Warrener P, et al. Complete genome sequence of Pseudomonas aeruginosa PAO1, an opportunistic pathogen. *Nature* 2000;**406**:959–64.
2. Rahme LG, Stevens EJ, Wolfort SF, Shao J, Tompkins RG, Ausubel FM. Common virulence factors for bacterial pathogenicity in plants and animals. *Science* 1995;**268**:1899–902.
3. Verma A, Arora SK, Kuravi SK, Ramphal R. Roles of specific amino acids in the N terminus of Pseudomonas aeruginosa flagellin and of flagellin glycosylation in the innate immune response. *Infect Immun* 2005;**73**:8237–46.
4. Inaba S, Hashimoto M, Jyot J, Aizawa S-I. Exchangeability of the flagellin (FliC) and the cap protein (FliD) among different species in flagellar assembly. *Biopolymers* 2013;**99**:63–72.
5. Tammam S, Sampaleanu LM, Koo J, Manoharan K, Daubaras M, Burrows LL, et al. PilMNOPQ from the Pseudomonas aeruginosa type IV pilus system form a transenvelope protein interaction network that interacts with PilA. *J Bacteriol* 2013;**195**:2126–35.
6. Merz AJ, So M, Sheetz MP. Pilus retraction powers bacterial twitching motility. *Nature* 2000;**407**:98–102.
7. Whitchurch CB, Alm RA, Mattick JS. The alginate regulator AlgR and an associated sensor FimS are required for twitching motility in Pseudomonas aeruginosa. *Proc Natl Acad Sci USA* 1996;**93**:9839–43.

8. Galle M, Carpentier I, Beyaert R. Structure and function of the Type III secretion system of Pseudomonas aeruginosa. *Curr Protein Pept Sci* 2012;**13**(8):831–42.
9. Patankar YR, Lovewell RR, Poynter ME, Jyot J, Kazmierczak BI, Berwin B. Flagellar motility is a key determinant of the magnitude of the inflammasome response to Pseudomonas aeruginosa. *Infect Immun* 2013;**81**(6):2043–52.

CHAPTER 20. RALSTONIA SOLANACEARUM

1. Genin S. Molecular traits controlling host range and adaptation to plants in Ralstonia solanacearum. *New Phytol* 2010;**187**(4):920–8.
2. Yao J, Allen C. Chemotaxis is required for virulence and competitive fitness of the bacterial wilt pathogen Ralstonia solanacearum. *J Bacteriol* 2006;**188**(10):3697–708.
3. Tans-Kersten J, Huang H, Allen C. Ralstonia solanacearum needs motility for invasive virulence on tomato. *J Bacteriol* 2001;**183**(12):3597–605.
4. Yao J, Allen C. The plant pathogen Ralstonia solanacearum needs aerotaxis for normal biofilm formation and interactions with its tomato host. *J Bacteriol* 2007;**189**(17):6415–24.
5. Salanoubat M, Genin S, Artiguenave F, Gouzy J, Mangenot S, Arlat M, et al. Genome sequence of the plant pathogen Ralstonia solanacearum. *Nature* 2002;**415**:497–502.
6. Tans-Kersten J, Brown D, Allen C. Swimming motility, a virulence trait of Ralstonia solanacearum, is regulated by FlhDC and the plant host environment. *Mol Plant Microbe Interact* 2004;**17**:686–95.
7. Pfund C, Tans-Kersten J, Dunning FM, Alonso JM, Ecker JR, Allen C, et al. Flagellin is not a major defense elicitor in Ralstonia solanacearum cells or extracts applied to Arabidopsis thaliana. *Mol Plant Microbe Interact* 2004;**17**:696–706.
8. Van Gijsegem F, Vasse J, Camus JC, Marenda M, Boucher C. Ralstonia solanacearum produces hrp-dependent pili that are required for PopA secretion but not for attachment of bacteria to plant cells. *Mol Microbiol* 2000;**36**(2):249–60.
9. Van Gijsegem F, Vasse J, De Rycke R, Castello P, Boucher C. Genetic dissection of Ralstonia solanacearum hrp gene cluster reveals that the HrpV and HrpX proteins are required for Hrp pilus assembly. *Mol Microbiol* 2002;**44**(4):935–46.

CHAPTER 21. RHODOBACTER SPHAEROIDES

1. Kobayashi K, Saitoh T, Shah DSH, Ohnishi K, Goodfellow IG, Sockett RE, et al. Purification and characterization of the flagellar basal body of Rhodobacter sphaeroides. *J Bacteriol* 2003;**185**:5295–300.
2. Castillo DJ, Ballado T, Camarena L, Dreyfus G. Functional analysis of a large non-conserved region of FlgK (HAP1) from Rhodobacter sphaeroides. *Antonie Van Leeuwenhoek* 2009;**95**:77–90.
3. Martínez-del Campo A, Ballado T, Camarena L, Dreyfus G. In Rhodobacter sphaeroides, chemotactic operon 1 regulates rotation of the flagellar system 2. *J Bacteriol* 2011;**193**:6781–6.

CHAPTER 22. RUEGERIA SP.

1. Biebl H, Allgaier M, Tindall BJ, Koblizek M, Lünsdorf H, Pukall R, et al. Dinoroseobacter shibae gen. nov., sp. nov., a new aerobic phototrophic bacterium isolated from dinoflagellates. *Int J Syst Evol Microbiol* 2005;**55**(Pt 3):1089–96.
2. Belas R, Horikawa E, Aizawa S-I, Suvanasuthi R. Genetic determinants of silicibacter sp. TM1040 motility. *J Bacteriol* 2009;**191**:4502–12.

CHAPTER 23. SACCHAROPHAGUS DEGRADANS

1. Weiner RM, Taylor II LE, Henrissat B, Hauser L, Land M, Coutinho PM, et al. Complete genome sequence of the complex carbohydrate-degrading marine bacterium, Saccharophagus degradans strain 2-40 T. *PLoS Genet* 2008;**4**:e1000087.
2. Shieh WY, Liu TY, Lin SY, Jean WD, Chen JS. Simiduia agarivorans gen. nov., sp. nov., a marine, agarolytic bacterium isolated from shallow coastal water from Keelung, Taiwan. *Int J Syst Evol Microbiol* 2008;**58**:895–900.

CHAPTER 24. SALMONELLA ENTERICA

1. Asakura S, Eguchi G, Iino T. Reconstitution of bacterial flagella in vitro. *J Mol Biol* 1964;**10**:42–56.
2. Asakura S, Eguchi G, Iino T. Unidirectional growth of Salmonella flagella in vitro. *J Mol Biol* 1968;**35**:227–36.
3. Asakura S. Polymerization of flagellin and polymorphism of flagella. *Advan In Biophs* 1970;**1**:99–155.
4. Iino T. Polarity of flagellar growth in Salmonella. *J Gen Microbiol* 1969;**56**:227–39.
5. Inaba S, Hashimoto M, Jyot J, Aizawa S-I. Exchangeability of the flagellin (FliC) and the cap protein (FliD) among different species in flagellar assembly. *Biopolymers* 2013;**99**:63–72.
6. Mizusaki H, Takaya A, Yamamoto T, Aizawa S-I. Signal pathway in the salt-activated expression of the SPI1/ type III secretion system in *Salmonella enterica* serovar Typhimurium. *J Bacteriol* 2008;**190**:4624–31.

For others: see References of Introduction

CHAPTER 25. SELENOMONAS RUMINANTIUM

1. Chalcroft JP, Bullivant S, Howard BH. Ultrastructural studies on Selenomonas ruminantium from the sheep rumen. *J Gen Microbiol* 1973;**79**:135–46.
2. Haya S, Tokumaru Y, Abe N, Kaneko J, Aizawa S-I. Characterization of lateral flagella in *Selenomonas ruminantium*. *Appl Environ Microbiol* 2011;**77**:2799–802.

CHAPTER 26. SINORHIZOBIUM MELILOTI

1. Rotter C, Mühlbacher S, Salamon D, Schmitt R, Scharf B. Rem, a new transcriptional activator of motility and chemotaxis in Sinorhizobium meliloti. *J Bacteriol* 2006;**188**(19):6932–42.
2. Trachtenberg S, DeRosier DJ, Aizawa S, Macnab RM. Pairwise perturbation of flagellin subunits. The structural basis for the differences between plain and complex bacterial flagellar filaments. *J Mol Biol* 1986;**190**:569–76.
3. Trachtenberg S, DeRosier DJ, Macnab RM. Three-dimensional structure of the complex flagellar filament of Rhizobium lupini and its relation to the structure of the plain filament. *J Mol Biol* 1987;**195**:603–20.
4. Sourjik V, Muschler P, Scharf B, Schmitt R. VisN and VisR are global regulators of chemotaxis, flagellar, and motility genes in *Sinorhizobium (Rhizobium) meliloti*. *J Bacteriol* 2000;**182**:782–8.
5. Scharf B, Schuster-Wolff-Bühring H, Rachel R, Schmitt R. Mutational analysis of the *Rhizobium lupnini* H13-3 and *Sinorhizobium meliloti flagellin genes: importance of flagellin A for flagellar* filament structure and transcriptional regulation. *J Bacteriol* 2001;**183**:5334–42.

CHAPTER 27. SYMBIOBACTERIUM THERMOPHILUM

1. Ueda K, Ohno M, Yamamoto K, Nara H, Mori Y, Shimada M, et al. Distribution and diversity of symbiotic thermophiles, Symbiobacterium thermophilum and related bacteria, in natural environments. *Appl Environ Microbiol* 2001;**67**(9):3779–84.
2. Ueda K, Yamashita A, Ishikawa J, Shimada M, Watsuji TO, Morimura K, et al. Genome sequence of Symbiobacterium thermophilum, an uncultivable bacterium that depends on microbial commensalism. *Nucleic Acids Res* 2004;**32**(16):4937–44.

CHAPTER 28. VIBRIO PARAHAEMOLITYCUSS

1. McCarter LL. Dual flagellar systems enable motility under different circumstances. *J Mol Microbiol Biotechnol* 2004;**7**:18–29.
2. Stewart BJ, McCarter LL. Lateral flagellar gene system of Vibrio parahaemolyticus. *J Bacteriol* 2003;**185**:4508–18.

3. Kim YK, McCarter LL. Analysis of the polar flagellar gene system of Vibrio parahaemolyti-cus. *J Bacteriol* 2000;**182**:3693–704.
4. McCarter LL. Polar flagellar motility of the Vibrionaceae. *Microbiol Mol Biol Rev* 2001;**65**:445–62.
5. Li N, Kojima S, Homma M. Sodium-driven motor of the polar flagellum in marine bacteria Vibrio. *Genes Cells* 2011;**16**:985–99.
6. Okabe M, Yakushi T, Homma M. Interactions of MotX with MotY and with the PomA/PomB sodium ion channel complex of the Vibrio alginolyticus polar flagellum. *J Biol Chem* 2005;**280**:25659–64.
7. Terashima H, Koike M, Kojima S, Homma M. The flagellar basal body-associated protein FlgT is essential for a novel ring structure in the sodium-driven Vibrio motor. J McCarter L. Regulation of flagella. *Curr Opin Microbiol* 2006;**9**:180–6.
8. Terashima H, Li N, Sakuma M, Koike M, Kojima S, Homma M, et al. Insight into the assembly mechanism in the supramolecular rings of the sodium-driven Vibrio flagellar motor from the structure of FlgT. *Proc Natl Acad Sci USA* 2013;**110**:6133–8.

CHAPTER 29. XANTHOMONAS ORYZAE

1. Lee BM, Park YJ, Park DS, Kang HW, Kim JG, Song ES, et al. The genome sequence of Xanthomonas oryzae pathovar oryzae KACC10331, the bacterial blight pathogen of rice. *Nucleic Acids Res* 2005;**33**:577–86.
2. Niño-Liu DO, Ronald PC, Bogdanove AJ. Xanthomonas oryzae pathovars: model pathogens of a model crop. *Mol Plant Pathol* 2006;**7**(5):303–24.
3. González JF, Myers MP, Venturi V. The inter-kingdom solo OryR regulator of Xanthomonas oryzae is important for motility. *Mol Plant Pathol* 2013;**14**(3):211–21.

CHAPTER 30. UNCHARACTERIZED SPECIES

1. Hattori R, Watanabe H, Tonosaka A, Hattori T. Unusual morphology of Agromonas oligotro-phica and the effect of NaCl and organice nutrient on its fine structure. *J Gen Appl Microbiol* 1995;**41**:23–30.
2. Mitsui H, Gorlach K, Lee H, Hattori R, Hattori T. Incubation time and media requirements of culturable bacteria from different phylogenetic groups. *J Microbiol Methods* 1997;**30**:103–10.
3. Saito A, Mitsui H, Hattori R, Minamisawa K. Slow-growing and oligotrophic soil bacteria phylogenetically close to Bradyrhizobium japonicum FEMS. *Microbiol Ecol* 1998;**25**:277–86.
4. Okubo T, Fukushima S, Itakura M, Oshima K, Longtonglang A, Teaumroong N, et al. Genome analysis suggests that the soil oligotrophic bacterium Agromonas oligotrophica (Bradyrhizobium oligotrophicum) is a nitrogen-fixing symbiont of Aeschynomene indica. *Appl Environ Microbiol* 2013;**79**:2542–51.
5. Ramírez-Bahena M-H, Chahboune R, Peix A, Velázquez E. Reclassification of Agromonas oligotrophica into the genus Bradyrhizobium as Bradyrhizobium oligotrophicum comb. nov. *Int J Syst Environ Microbiol* 2013;**63**:1013–6.

CHAPTER 31. BUCHNERA APHIDICOLA

1. Shigenobu S, Watanabe H, Hattori M, Sakaki Y, Ishikawa H. Genome sequence of the endo-cellular bacterial symbiont of aphids Buchnera sp. APS. *Nature* 2000;**407**:81–6.
2. Maezawa K, Shigenobu S, Taniguchi H, Kubo T, Aizawa S-I, Morioka M. Hundreds of the fla-gellar basal bodies cover the cell surface of the endosymbiotic bacterium Buchnera aphidicola sp. APS. *J Bacteriol* 2006;**188**:6539–43.

CHAPTER 32. METHANOCOCCUS VOLTAE

1. Jarrel KF, Bayley DP, Kostyukova AS. The archaeal flagellum: a unique motility structure. *J Bacteriol* 1996;**178**:5057–64.

2. Bardy SL, Mori T, Komoriya K, Aizawa S-I, Jarrell KF. Identification and localization of flagellins FlaA and FlaB3 within the flagella of *Methanococcus voltae*. *J Bacteriol* 2002;**184**:5223–33.
3. Jarrell KF, Albers SV. The archaellum: an old motility structure with a new name. *Trends Microbiol* 2012;**20**:307–12.

CHAPTER 33. MYXOCOCCUS XANTHUS

1. Spormann AM. Gliding motility in bacteria: insights from studies of Myxococcus xanthus. *Microbiol Mol Biol Rev* 1999;**63**(3):621–41.
2. Mauriello EM, Mignot T, Yang Z, Zusman DR. Gliding motility revisited: how do the myxobacteria move without flagella? *Microbiol Mol Biol Rev* 2010;**74**:229–49.
3. Nan B, Chen J, Neu JC, Berry RM, Oster G, Zusman DR. Myxobacteria gliding motility requires cytoskeleton rotation powered by proton motive force. *Proc Natl Acad Sci USA* 2011;**108**:2498–503.
4. Nan B, Bandaria JN, Moghtaderi A, Sun IH, Yildiz A, Zusman DR. Flagella stator homologs function as motors for myxobacterial gliding motility by moving in helical trajectories. *Proc Natl Acad Sci USA* 2013;**110**:E1508–13.

CHAPTER 34. SAPROSPIRA GRANDIS

1. Lewin RA. *Saprospira grandis*: a flexibacterium that can catch bacterial prey by ixotrophy. *Microb Ecol* 1997;**34**:232–6.
2. Lewin RA. Rod-shaped particles in *Saprospira*. *Nature* 1963;**198**:103–4.
3. Saw JHW, Yuryev A, Kanbe M, Hou S, Young AG, Aizawa S-I, et al. Complete genome sequencing and analysis of *Saprospira grandis* str. Lewin, a predatory marine bacterium. *Stand Genomic Sci* 2012;**6**:84–93.

CHAPTER 35. SHIGELLA FLEXNERI

1. Tamano K, Aizawa S, Katayama E, Nonaka T, Imajoh-Ohmi S, Kuwae A, et al. Supramolecular structure of Shigella type III secretion machinery: the needle part is changeable in length and essential for delivery of effectors. *EMBO J* 2000;**19**:3876–87.
2. Tominaga A, Mahmoud MA, Mukaihara T, Enomoto M. Molecular characterization of intact, but cryptic, flagellin genes in the genus Shigella. *Mol Microbiol* 1994;**12**:277–85.
3. Tominaga A, Lan R, Reeves PR. Evolutionary changes of the flhDC flagellar master operon in Shigella strains. *J Bacteriol* 2005;**187**:4295–302.
4. Girón JA. Expression of flagella and motility by Shigella. *Mol Microbiol* 1995;**18**:63–75.

Topic 1: Gene regulation

1. Yanagihara S, Iyoda S, Ohnishi K, Iino T, Kutsukake K. Structure and transcriptional control of the flagellar master operon of *Salmonella typhimurium*. *Genes Genet Syst* 1999;**74**(3):105–11.
2. Lee YY, Barker CS, Matsumura P, Belas R. Refining the binding of the *Escherichia coli* flagellar master regulator, FlhD4C2, on a base-specific level. *J Bacteriol* 2011;**193**(16):4057–68.
3. Claret L, Hughes C. Interaction of the atypical prokaryotic transcription activator FlhD2C2 with early promoters of the flagellar gene hierarchy. *J Mol Biol* 2002;**321**(2):185–99.
4. Shi W, Zhou Y, Wild J, Adler J, Gross CA. DnaK, DnaJ, and GrpE are required for flagellum synthesis in Escherichia coli. *J Bacteriol* 1992;**174**(19):6256–63.
5. Shin S, Park C. Modulation of flagellar expression in Escherichia coli by acetyl phosphate and the osmoregulator OmpR. *J Bacteriol* 1995;**177**(16):4696–702.
6. Ikebe T, Iyoda S, Kutsukake K. Promoter analysis of the class 2 flagellar operons of Salmonella. *Genes Genet Syst* 1999;**74**(4):179–83.
7. Kutsukake K, Ikebe T, Yamamoto S. Two novel regulatory genes, fliT and fliZ, in the flagellar regulon of Salmonella. *Genes Genet Syst* 1999;**74**(6):287–92.

8. Kutsukake K, Iino T. Role of the FliA FlgM regulator system on the transcriptional control of the flagellar regulon and flagellar formation in Salmonella typhimurium. *J Bacteriol* 1994;**176**:3598–605.
9. Chadsey MS, Karlinsey JE, Hughes KT. The flagellar anti-sigma factor FlgM actively dissociates Salmonella typhimurium sigma28 RNA polymerase holoenzyme. *Genes Dev* 1998;**12**(19):3123–36.
10. Kutsukake K. Excretion of the anti-sigma factor through a flagellar substructure couples flagellar gene expression with flagellar assembly in Salmonella typhimurium. *Mol Gen Genet* 1994;**243**(6):605–12.
11. Karlinsey JE, Tanaka S, Bettenworth V, Yamaguchi S, Boos W, Aizawa SI, et al. Completion of the hook-basal body complex of the Salmonella typhimurium flagellum is coupled to FlgM secretion and fliC transcription. *Mol Microbiol* 2000;**37**(5):1220–31.
12. Wozniak CE, Chevance FF, Hughes KT. Multiple promoters contribute to swarming and the coordination of transcription with flagellar assembly in Salmonella. *J Bacteriol* 2010;**192**:4752–62.
13. Dasgupta N, Wolfgang MC, Goodman AL, Arora SK, Jyot J, Lory S, et al. A four-tiered transcriptional regulatory circuit controls flagellar biogenesis in Pseudomonas aeruginosa. *Mol Microbiol* 2003;**50**:809–24.
14. Arora SK, Ritchings BW, Almira EC, Lory S, Ramphal R. A transcriptional activator, FleQ, regulates mucin adhesion and flagellar gene expression in Pseudomonas aeruginosa in a cascade manner. *J Bacteriol* 1997;**179**:5574–81.

Topic 2: Gene arrangement

1. Gupta RS. Protein phylogenies and signature sequences: a reappraisal of evolutionary relationships among archaebacteria, eubacteria, and eukaryotes. *Microbiol Mol Biol Rev* 1998;**62**(4):1435–91.
2. Gupta RS. The phylogeny of proteobacteria: relationships to other eubacterial phyla and eukaryotes. *FEMS Microbiol Rev* 2000;**24**:367–402.
3. Kunisawa T. Gene arrangements and phylogeny in the class Proteobacteria. *J Theor Biol* 2001;**213**:9–19.

Topic 3: Mot proteins

1. Braun TF, Al-Mawsawi LQ, Kojima S, Blair DF. Arrangement of core membrane segments in the MotA/MotB proton-channel complex of Escherichia coli. *Biochemistry* 2004;**243**:35–45.
2. Leake MC, Chandler JH, Wadhams GH, Bai F, Berry RM, Armitage JP. Stoichiometry and turnover in single, functioning membrane protein complexes. *Nature* 2006;**443**(7109):355–8.
3. Zhou J, Blair DF. Residues of the cytoplasmic domain of MotA essential for torque generation in the bacterial flagellar motor. *J Mol Biol* 1997;**273**(2):428–39.
4. Lloyd SA, Whitby FG, Blair DF, Hill CP. Structure of the C-terminal domain of FliG, a component of the rotor in the bacterial flagellar motor. *Nature* 1999;**400**(6743):472–5.
5. Roujeinikova A. Crystal structure of the cell wall anchor domain of MotB, a stator component of the bacterial flagellar motor: implications for peptidoglycan recognition. *Proc Natl Acad Sci USA* 2008;**105**(30):10348–53.
6. Sudo Y, Kitade Y, Furutani Y, Kojima M, Kojima S, Homma M, et al. Interaction between Na+ ion and carboxylates of the PomA-PomB stator unit studied by ATR-TIR spectroscopy. *Biochemistry* 2009;**48**(49):11699–705.
7. Takekawa N, Terauchi T, Morimoto YV, Minamino T, Lo CJ, Kojima S, et al. Na+ conductivity of the Na+-driven flagellar motor complex composed of unplugged wild-type or mutant PomB with PomA. *J Biochem* 2013;**153**(5):44151.
8. Ito M, Hicks DB, Henkin TM, Guffanti AA, Powers B, et al. MotPS is the stator-force generator for motlty of alkaliphilic Bacillus and its homologue is a second functional Mot in Bacillus subtilis. *Mol Microbol* 2004;**53**:1035–49.
9. Ito M, Terahara N, Fujinami S, Krulwich TA. Properties of motility in Bacillus subtilis powered by the H+-coupled MotAB flagellar stator, Na+-coupled MotPS or hybrid stators MotAS or MotPB. *J Mol Biol* 2005;**352**:396–408.
10. Terahara N, Sano M, Ito M. A Bacillus flagellar motor that can use both Na+ and K+ as a coupling ion is converted by a single mutation to use only Na+. *PLoS One* 2012;**7**:e46248.

11. Braun TF, Blair DF. Targeted disulfide cross-linking of the MotB protein of Escherichia coli: evidence for two H(+) channels in the stator Complex. *Biochemistry* 2001;**40**:13051–9.
12. Asai Y, Kawagishi I, Sockett RE, Homma M. Coupling ion specificity of chimeras between H+- and Na+-driven motor proteins, MotB and PomB, in Vibrio polar flagella. *EMBO J* 2000;**19**:3639–48.
13. Okabe M, Yakushi T, Kojima M, Homma M. MotX and MotY, specific components of the sodium-driven flagellar motor, colocalize to the outer membrane in Vibrio alginolyticus. *Mol Microbiol* 2002;**46**(1):125–34.
14. Okabe M, Yakushi T, Homma M. Interactions of MotX with MotY and with the PomA/PomB sodium ion channel complex of the Vibrio alginolyticus polar flagellum. *J Biol Chem* 2005;**280**(27):25659–64.
15. Terashima H, Fukuoka H, Yakushi T, Kojima S, Homma M. The Vibrio motor proteins, MotX and MotY, are associated with the basal body of Na-driven flagella and required for stator formation. *Mol Microbiol* 2006;**62**:1170–80.
16. Doyle TB, Hawkins AC, McCarter LL. The complex flagellar torque generator of Pseudomonas aeruginosa. *J Bacteriol* 2004;**186**(19):6341–50.

Topic 4: Flagellin size

1. Yonekura K, Maki-Yonekura S, Namba K. Complete atomic model of the bacterial flagellar filament by electron cryomicroscopy. *Nature* 2003;**424**(6949):643–50.
2. Yoshioka K, Aizawa S, Yamaguchi S. Flagellar filament structure and cell motility of salmonella typhimurium mutants lacking part of the outer domain of flagellin. *J Bacteriol* 1995;**177**:1090–3.
3. Smith NH, Selander RK. Sequence invariance of the antigen-coding central regions of the phase I flagellar filament gene (fliC) among strains of Salmonella typhimurium. *J Bacteriol* 1990;**172**:603–9.
4. Hayashi F, Smith KD, Ozinsky A, Hawn TR, Yi EC, Goodlett DR, et al. The innate immune response to bacterial flagellin is mediated by Toll-like receptor 5. *Nature* 2001;**410**:1099–103.
5. Murthy KG, Deb A, Goonesekera S, Szabó C, Salzman AL. Identification of conserved domains in Salmonella muenchen flagellin that are essential for its ability to activate TLR5 and to induce an inflammatory response in vitro. *J Biol Chem* 2004;**279**(7):5667–75.
6. Newton SM, Jacob CO, Stocker BA. Immune response to cholera toxin epitope inserted in Salmonella flagella. *Science* 1989;**244**(4900):70–2.
7. Tanskanen J, Korhonen TK, Westerlund-Wikström B. Construction of a multihybrid display system, flagellar filaments carrying two foreign adhesive peptides. *Appl Environ Microbiol* 2000;**66**:4152–6.
8. Woods RD, Takahashi N, Aslam A, Pleass RJ, Aizawa S, Sockett RE. Bifunctional nanotube scaffolds for diverse ligands are purified simply from Escherichia coli strains coexpressing two functionalized flagellar genes. *Nano Lett* 2007;**7**:1809–16.
9. Szabó V, Muskotál A, Tóth B, Mihovilovic MD, Vonderviszt F. Construction of a xylanase a variant capable of polymerization. *PLoS One* 2011;**6**:e25388.
10. Klein Á, Tóth B, Jankovics H, Muskotál A, Vonderviszt F. A polymerizable GFP variant. *Protein Eng Des Sel* 2012;**25**:153–7.
11. Singer HM, Erhardt M, Steiner AM, Zhang MM, Yoshikami D, Bulaj G, et al. Selective purification of recombinant neuroactive peptides using the flagellar type III secretion system. *MBio* 2012;**3**(3). pii: e00115–12.

Topic 5: Flagella and Pathogenicity

1. Richardson K. Roles of motility and flagellar structure in pathogenicity of Vibrio cholerae: analysis of motility mutants in three animal models. *Infect Immun* 1991;**59**(8):2727–36.
2. Drake D, Montie TC. Flagella, motility and invasive virulence of Pseudomonas aeruginosa. *J Gen Microbiol* 1988;**134**(1):43–52.
3. Mulholland V, Hinton JC, Sidebotham J, Toth IK, Hyman LJ, Pérombelon MC, et al. A pleiotropic reduced virulence (Rvi-) mutant of Erwinia carotovora subspecies atroseptica is defective in flagella assembly proteins that are conserved in plant and animal bacterial pathogens. *Mol Microbiol* 1993;**9**(2):343–56.

4. Akerley BJ, Cotter PA, Miller JF. Ectopic expression of the flagellar regulon alters development of the Bordetella-host interaction. *Cell* 1995;**80**(4):611–20.
5. Chesnokova O, Coutinho JB, Khan IH, Mikhail MS, Kado CI. Characterization of flagella genes of Agrobacterium tumefaciens, and the effect of a bald strain on virulence. *Mol Microbiol* 1997;**23**(3):579–90.
6. Tans-Kersten J, Huang H, Allen C. Ralstonia solanacearum needs motility for invasive virulence on tomato. *J Bacteriol* 2001;**183**(12):3597–605.
7. Chua KL, Chan YY, Gan YH. Flagella are virulence determinants of Burkholderia pseudomallei. *Infect Immun* 2003;**71**(4):1622–9.
8. O'Neil HS, Marquis H. Listeria monocytogenes flagella are used for motility, not as adhesins, to increase host cell invasion. *Infect Immun* 2006;**74**(12):6675–81.
9. Chubiz JE, Golubeva YA, Lin D, Miller LD, Slauch JM. FliZ regulates expression of the Salmonella pathogenicity island 1 invasion locus by controlling HilD protein activity in Salmonella enterica serovar typhimurium. *J Bacteriol* 2010;**92**:6261–70.
10. Hung CC, Haines L, Altier C. The flagellar regulator fliT represses Salmonella pathogenicity island 1 through flhDC and fliZ. *PLoS One* 2012;**7**:e34220.
11. Kubori T, Matsushima Y, Nakamura D, Uralil J, Lara-Tejero M, Sukhan A, et al. Supramolecular structure of the *Salmonella typhimurium* type III protein secretion system. *Science* 1998;**280**:602–5.
12. Galkin VE, Schmied WH, Schraidt O, Marlovits TC, Egelman EH. The structure of the Salmonella typhimurium type III secretion system needle shows divergence from the flagellar system. *J Mol Biol* 2010;**396**(5):1392–7.
13. Hueck CJ. Type III protein secretion systems in Bacterial pathogens of animals and plants. *Microbiol & Mol Biol Rev* 1998;**62**:379–433.

Topic 6: Flagellar position and shape

1. Leifson E. *Atlas of bacterial flagellation*. New York: Academic Press; 1960. 171 pp.
2. Haya S, Tokumaru Y, Abe N, Kaneko J, Aizawa S-I. Characterization of lateral flagella in Selenomonas ruminantium. *Appl Environ Microbiol* 2011;**77**:2799–802.
3. McCarter LL. Dual flagellar systems enable motility under different circumstances. *J Mol Microbiol Biotechnol* 2004;**7**:18–29.
4. Kusumoto A, Shinohara A, Terashima H, Kojima S, Yakushi T, Homma M. Collaboration of FlhF and FlhG to regulate polar-flagella number and localization in Vibrio alginolyticus. *Microbiology* 2008;**154**(Pt 5):1390–9.
5. Armitage JP, Macnab RM. Unidirectional, intermittent rotation of the flagellum of Rhodobacter sphaeroides. *J Bacteriol* 1987;**169**(2):514–8.
6. Kazmierczak BI, Hendrixson DR. Spatial and numerical regulation of flagellar biosynthesis in polarly flagellated bacteria. *Mol Microbiol* 2013;**88**:655–63.
7. Dasgupta N, Arora SK, Ramphal R. fleN, a gene that regulates flagellar number in Pseudomonas aeruginosa. *J Bacteriol* 2000;**182**:357–64.
8. Schniederberend M, Abdurachim K, Murray TS, Kazmierczak BI. The GTPase activity of FlhF is dispensable for flagellar localization, but not motility, in Pseudomonas aeruginosa. *J Bacteriol* 2013;**195**:1051–60.
9. Kojima M, Nishioka N, Kusumoto A, Yagasaki J, Fukuda T, Homma M. Conversion of monopolar to peritrichous flagellation in Vibrio alginolyticus. *Microbiol Immunol* 2011;**55**(2):76–83.
10. Guttenplan SB, Shaw S, Kearns DB. The cell biology of peritrichous flagella in Bacillus subtilis. *Mol Microbiol* 2013;**87**:211–29.
11. Fujii M, Shibata S, Aizawa S-I. Polar, peritrichous, and lateral flagella belong to three distinguishable flagellar families. *J Mol Biol* 2008;**379**:273–83.

Topic 7: History of *Salmonella* SJW strain

1. Lilleengen K. Typing *Salmonella typhimurium* by means of bacteriophage. *Acta Pathol Microbiol Scand Supple* 1948;**77**:11–125.
2. Zinder ND, Lederberg J. Genetic exchange in Salmonella. *J Bacteriol* 1952;**64**:679–99.
3. Stocker BAD, Zinder ND, Lederberg J. Transduction of flagellar characters in Salmonella. *J Gen Microbiol* 1953;**9**:410–33.

Topic 8: Hook length

1. Silverman MR, Simon MI. Flagellar assembly mutants in *Escherichia coli*. *J Bacteriol* 1972;**112**:986–93.
2. Patterson-Delafield J, Martinez RJ, Stocker BA, Yamaguchi S. A new fla gene in *Salmonella typhimurium*–*flaR*–and its mutant phenotype-superhooks. *Arch Mikrobiol* 1973;**90**:107–20.
3. Hirano T, Yamaguchi S, Oosawa K, Aizawa S-I. Roles of FliK and FlhB in determination of flagellar hook length in *Salmonella typhimurium*. *J Bacteriol* 1994;**176**:5439–49.
4. Journet L, Agrain C, Broz P, Cornelis GR. The needle length of bacterial injectisomes is determined by a molecular ruler. *Science* 2003;**302**:1757–60.
5. Wagner S, Stenta M, Metzger LC, Dal Peraro M, Cornelis GR. Length control of the injectisome needle requires only one molecule of Yop secretion protein P (YscP). *Proc Natl Acad Sci USA* 2010;**107**:13860–5.
6. Minamino T, Pugsley AP. Measure for measure in the control of type III secretion hook and needle length. *Mol Microbiol* 2005;**56**:303–8.
7. Haya S, Tokumaru Y, Abe N, Kaneko J, Aizawa S-I. Characterization of lateral flagella in *Selenomonas ruminantium*. *Appl Environ Microbiol* 2011;**77**:2799–802.
8. Shibata S, Takahashi N, Chevance FF, Karlinsey JE, Hughes KT, Aizawa SI. FliK regulates flagellar hook length as an internal ruler. *Mol Microbiol* 2007;**64**:1404–15.
9. Agrain C, Callebaut I, Journet L, Sorg I, Paroz C, Mota LJ, et al. Characterization of a Type III secretion substrate specificity switch (T3S4) domain in YscP from Yersinia enterocolitica. *Mol Microbiol* 2005;**56**(1):54–67.
10. Kubori T, Matsushima Y, Nakamura D, Uralil J, Lara-Tejero M, Sukhan A, et al. Supramolecular structure of the *Salmonella typhimurium* type III protein secretion system. *Science* 1998;**280**:602–5.
11. Tamano K, Aizawa S-I, Katayama E, Nonaka T, Imajoh-Ohmi S, Kuwae A, et al. Supramolecular structure of the *Shigella* type III secretion machinery: the needle part is changeable in length and essential for delivery of effectors. *EMBO J* 2000;**19**:3876–87.
12. Uchida K, Dono K, Aizawa S. Length control of the flagellar hook in a temperature-sensitive flgE mutant of salmonella enterica serovar typhimurium. *J Bacteriol* 2013;**195**(16):3590–5.
13. Aizawa S-I. Mystery of FliK in length control of the flagellar hook. *J Bacteriol* 2012;**194**:4798–800.
14. Aizawa S-I. Rebuttal: flagellar hook length is controlled by a secreted molecular ruler. *J Bacteriol* 2012;**194**:4797.
15. Hughes KT. Flagellar hook length is controlled by a secreted molecular ruler. *J Bacteriol* 2012;**194**:4793–6.
16. Hughes KT. Rebuttal: mystery of FliK in length control of the flagellar hook. *J Bacteriol* 2012;**194**:4801.
17. Mizuno S, Amida H, Kobayashi N, Aizawa S-I, Tate S. The NMR structure of FliK, the trigger for the switch of substrate specificity in the flagellar type III secretion apparatus. *J Mol Biol* 2011;**409**:558–73.

Topic 9: Multiple flagellins

1. Driks A, Bryan R, Shapiro L, DeRosier DJ. The organization of the Caulobacter crescentus flagellar filament. *J Mol Biol* 1989;**206**:627–36.
2. Iida Y, Hobley L, Lambert C, Sockett E, Aizawa S-I. Roles of multiple flagellins in flagellar formation and flagellar growth post bdelloplast- lysis, in *Bdellovibrio bacteriovorus*. *J Mol Biol* 2009;**394**:1011–21.
3. Faulds-Pain A, Birchall C, Aldridge C, Smith WD, Grimaldi G, Nakamura S, et al. Flagellin redundancy in Caulobacter crescentus and its implications for flagellar filament assembly. *J Bacteriol* 2011;**193**:2695–707.

APPENDIX
Flagellar family

1. Shimada K, Kamiya R, Asakura S. Left-handed to right-handed helix conversion in Salmonella flagella. *Nature* 1975;**254**:332–4.
2. Kamiya R, Hotani H, Asakura S. Polymorphic transition in bacterial flagella. In: Amos WB, Duckett JD, editors. *Prokaryotic and Eukaryotic Flagella*. Cambridge: Cambridge University-Press; 1982. p. 53–76.

3. Hotani H. Micro-video study of moving bacterial flagellar filaments. III. Cyclic transformation induced by mechanical force. *J Mol Biol* 1982;**156**:791–806.
4. Kanto S, Okino H, Aizawa S-I, Yamaguchi S. Amino acids responsible for flagellar shape are distributed in terminal regions of flagellin. *J Mol Biol* 1991;**219**:471–80.
5. Macnab RM, Ornston MK. Normal-to-curly flagellar transitions and their role in bacterial tumbling: stabilization of an alternative quaternary structure by mechanical force. *J Mol Biol* 1977;**112**:1–30.
6. Asakura S. Polymerization of flagellin and polymorphism of flagella. *Adv Biophs* 1970;**1**:99–155.
7. Calladine CR. Change of waveform in bacterial flagella: the role of mechanics at the molecular level. *J Mol Biol* 1978;**118**:457–79.
8. Hasegawa K, Yamashita I, Namba K. Quasi- and nonequivalence in the structure of bacterial flagellar filament. *Biophy J* 1998;**74**:569–75.
9. Shibata S, Alam M, Aizawa S-I. Flagella of the deep-sea bacteria *Idiomarina loihiensis* belong to a family different from *Salmonella* flagella. *J Mol Biol* 2005;**352**:510–6.
10. Fujii M, Shibata S, Aizawa S-I. Polar, peritrichous, and lateral flagella belong to three distinguishable flagellar families. *J Mol Biol* 2008;**379**:273–83.
11. Kusumoto A, Nishioka N, Kojima S, Homma M. Mutational analysis of the GTP-binding motif of FlhF which regulates the number and placement of the polar flagellum in Vibrio alginolyticus. *J Biochem* 2009;**146**(5):643–50.
12. Galkin VE, Yu X, Bielnicki J, Heuser J, Ewing CP, Guerry P, et al. Divergence of quaternary structures among bacterial flagellar filaments. *Science* 2008;**320**(5874):382–5.

Protocols for purification of flagella
1. Okino H, Isomura M, Yamaguchi S, Magariyama Y, Kudo S, Aizawa SI. Release of flagellar filament-hook-rod complex by a Salmonella typhimurium mutant defective in the M ring of the basal body. *J Bacteriol* 1989;**171**:2075–82.
2. Aizawa S-I, Dean GE, Jones CJ, Macnab RM, Yamaguchi S. Purification and characterization of the flagellar hook-basal body complex of Salmonella typhimurium. *J Bacteriol* 1985;**161**:836–49.
3. Mizusaki H, Takaya A, Yamamoto T, Aizawa S-I. Signal pathway in the salt-activated expression of the SPI1/ type III secretion system in *Salmonella enterica* serovar Typhimurium. *J Bacteriol* 2008;**190**:4624–31.

Microscopic techniques for observation of flagella
1. Leifson E. *Atlas of bacterial flagellation*. New York and London: Academic Press; 1969.
2. Macnab RM, Ornston MK. Normal-to-curly flagellar transitions and their role in bacterial tumbling: stabilization of an alternative quaternary structure by mechanical force. *J Mol Biol* 1977;**112**:1–30.
3. Hotani H. Micro-video study of moving bacterial flagellar filaments. III. Cyclic transformation induced by mechanical force. *J Mol Biol* 1982;**156**:791–806.
4. Turner L, Ryu WS, Berg HC. Real-time imaging of fluorescent flagellar filaments. *J Bacteriol* 2000;**182**:2793–801.
5. Turner L, Stern AS, Berg HC. Growth of flagellar filaments of Escherichia coli is independent of filament length. *J Bacteriol* 2012;**194**:2437–42.
6. Jonas K, Tomenius H, Kader A, Normark S, Römling U, Belova LM, et al. Roles of curli, cellulose and BapA in Salmonella biofilm morphology studied by atomic force microscopy. *BMC Microbiol* 2007;**7**:70.
7. Naya M, Mononobe S, Maheswari RM, Saiki T, Ohtsu M. Imaging of biological samples by a collection-mode photon scanning tunneling microscope with an apertured probe. *Opt Commun* 1996;**124**:9–15.
8. Naya M, Micheletto R, Mononobe S, Uma Maheswari R, Ohtsu M. Near-field optical imaging of flagellar filaments of salmonella in water with optical feedback control. *Appl Opt* 1997;**36**(7):1681–3.

Note: Page numbers followed by "*f*" and "*t*" refers to figures and tables respectively.

A

Acetic acid, 42
Acquired immune system, 54
Adhesion, 30, 62, 64–65
Adventurous (A) motility, 96
Aerotaxis, 64
Aflagellate, 98–99
Alfalfa, 82
Anti-sigma factor, 24–25, 65
Archaellum, 95
Asymmetrical, 32
Avirulence (Avr), 58
Axial protein, 4

B

Bacteriocytes, 92–93
Bdelloplast, 26–27
Biofilm, 64, 70, 96
Bioluminescence, 16–17
Bipolar flagella, 52–53
Blackleg, 58
Blight, 88
Bobtail squid, 16

C

Carbonic anhydrase, 84
Catabolite repression, 36, 78, 88
Chlorophylls, 66
Coculture, 84
Complex filament, 82–83
Complex polysaccharides, 72
Compost, 84
Cooperative motility, 56
Cryptic flagella, 39, 101

D

Diguanylate cyclase, 33

E

Effector, 58, 76, 100–101
Endoflagella, 28
Endosymbiotic, 92
Energy taxis, 18
Enteroadherent *E. coli* (EAEC), 38–39
EPEC, 38–39

EspA filament, 38–39
Exoenzyme, 58
Extracellular polysaccharide, 64

F

Fat body, 92
Fla 2, 67–68
Flag-2, 39
flgM, 24–25, 55, 65, 75
flgO, 87
flgT, 87
flhFG, 13, 23, 31, 47, 53, 69, 79, 85
fliK, 21, 24–25, 52, 80–81, 80*t*, 88–89
f-Met, 75
Fly-paper, 98
Foodborne illness, 86
Formate, 94
Fructose, 94

G

Gastric cancer, 44
Gastritis, 44
Gastroenteritis, 86, 100–101
Gene transfer, 37–38
Gliding, 96–99
Glycosylation, 21, 33, 45, 61

H

hag, 23, 41
Halophilic, 48
HAP region, 3, 9–10, 10*f*, 13, 68, 80
Heme, 66
High (G + C), 84
hil, 76
Hrp pili, 63*t*
hrp, 59, 65, 89
Hydrothermal, 48
Hyperflagellate, 58–59
Hyper-variable, 54

I

Infectivity, 44
Innate immune system, 54
Invasion, 76
Ixotrophy, 98

L
Lactic acid, 34
laf, 1, 19–20
Leaf streak, 88
Legume, 82
Linear chromosome, 29
Liquefaction, 72
Lithotrophy, 66
Low cell density, 83
lux, 16–17
Lyme disease, 28
Lysozyme, 34–35, 107–109

M
Macrophages, 76
Magnetite crystals, 52
Magnetotactic, 52
Mariana Trench, 40
Mega-plasmid, 65
Mesophilic, 94
Methane, 94
Microcolonies, 56
Minimal medium (MM), 30–31
Mobilizing plasmids, 61
Morphotype, 56
motP, 23
motS, 23, 47
MS ring complex, 4
Murein transglycosylase, 87
Mycelia, 15

N
Nebula-like colonies, 56
Needle complex, 11, 63, 76
Negative staining, 9–10, 113
Nitrogen-fixing, 18, 30–31, 66
Nodules, 30, 82–83

O
Opportunistic pathogen, 34, 50, 60

P
Pathogenic *E. coli*, 37–38, 37*t*, 101
Pea aphid, 92–93
Pectin, 58
Peptic ulcer, 44
Periplasmic flagella, 4, 28
Periplasmic space, 21
Phosphotungstic acid (PTA), 9–10
Photosynthetic, 66
Phytoplankton, 70
pilA, 62, 97

PL ring complex, 13, 22–23, 40, 75
Plain filament, 82–83
Plant pathogen, 64, 88
pleD, 33
pof, 1, 46
Polymorphs, 48–49, 102–105, 102*t*
pomAB, 47
Probiotics, 34
Prosthecae, 72
Proton-driven motor, 23, 27, 51, 53
Protozoa, 50
rem, 83

R
Rhapidosome, 99
Rhizosphere, 18, 82
Root hair, 30
Root nodules, 30
Roseobacter, 70
Rumen, 69, 78

S
Salmonella Pathogenicity Island (SPI), 63, 76
Salt-sensitive, 90–91
Sec general pathway, 4, 75
Septicemia, 34
Sessile cell, 32
Sheath, 15, 44
sigD, 23
Sigma factor 28, 24–25
Signal peptides, 4, 21, 75, 95
Slow growth, 90
Social (S) motility, 56, 96
Soft rot disease, 58
Soil extract, 14
Sphaeroplasts, 108
Sphingobacteria, 98
SPI1, 63, 76
SPI2, 76
SPI5, 76
Spirochete, 28, 46
Sporangium (a), 14
Spore, 14–15, 22
Stalk, 32–33
Stomach, 44
Submarine volcano, 48
Subpolar, 1, 30–32, 58–59, 69, 85, 104
Surface (S) layer, 30
Swimming speed, 14
Swim-or-stick, 70
Symbioses, 70
Syntrophic, 84

T

T ring, 87
Tapered, 81
Tetrapyrroles, 66
tgl, 97
Thermophilic, 40
Thick flagellum, 30–31
Thin flagellum, 30–31, 83
Ticks, 28
Toll-like receptor (TLR) 5, 54
Twitching, 62
TYC medium, 30–31
Type III secretion system (T3SS), 4, 21, 30, 58, 63, 75, 100–101
Type IV pili, 26–27, 62, 64, 96
Type IV secretion system (T4SS), 50–51

U

Uranyl acetate (UA), 9–10

V

Vacuoles, 72
Viable but nonculturable (VBNC), 90
Vinegar, 42
Viscosity, 108
visN, 83
Vitamin B12, 66

W

Wilt disease, 64

X

Xanthan, 88